Aircraft Engines and Gas Turbines

Aircraft Engines and Gas Turbines

Jack L. Kerrebrock

The MIT Press
Cambridge, Massachusetts, and London, England

Eighth Printing, 1989

Copyright © 1977 by
The Massachusetts Institute of Technology

Printed and bound in the United States of America by Halliday Lithograph.

Library of Congress Cataloging in Publication Data
Kerrebrock, Jack L
Aircraft engines and gas turbines.
Includes index.
1. Aircraft gas turbines. I. Title.
TL709.K46 629.134'353 77–4428
ISBN 0–262–11064–4

Contents

Preface ix

Acknowledgments xi

1
Introduction to Concepts 1

1.1 Thermal Efficiency 1
1.2 Propulsive Efficiency 2
1.3 Specific Impulse and Range 3
1.4 Ramjets 4
1.5 Turbojets 6
1.6 Turbofans 7
1.7 Shaft Engines: Regeneration 8
1.8 Stationary Gas Turbines: Topping 11
1.9 Energy Exchange, Mach Number, Reynolds Number 11
1.10 Stresses 13
1.11 Noise 13
1.12 Thrust and Drag 14
1.13 Some Engines in Cutaway 17
Problems 20

2
Ideal Cycle Analysis: Trends 21

2.1 Stagnation Temperature and Pressure 22
2.2 The Ramjet 22
2.3 The Turbojet 25
2.4 The Afterburning Turbojet 28
2.5 The Turbofan 31
2.6 The Afterburning Turbofan 34
2.7 The Turboprop 35
2.8 The Regenerative Gas Turbine 39
2.9 Gas Turbines for Topping 42
2.10 The Importance of Turbine Inlet Temperature 43
Problems 46

3
Quantitative Cycle Analysis 48

3.1 Variation in Gas Properties 48
3.2 Diffuser Pressure Recovery 49
3.3 Compressor and Turbine Efficiencies 50
3.4 Burner Efficiency and Pressure Loss 51
3.5 Imperfect Expansion Loss 52

3.6 Heat Exchanger Effectiveness and Pressure Loss 53
3.7 Turbojet with Losses 54
3.8 Regenerative Gas Turbine with Losses 57
3.9 Combined Gas Turbine–Steam Cycles with Losses 61
3.10 Concluding Comments 64
References 65
Problems 65

4
Nonrotating Components 66

4.1 Summary of Gas Dynamics 66
4.2 Diffusers 78
4.3 Exhaust Nozzles 93
4.4 Combustors and Afterburners 99
References 113
Problems 114

5
Compressors 116

5.1 Energy Exchange, Rotor to Fluid 116
5.2 Compressor Geometry and the Flow Pattern 123
5.3 Limits on Stage Pressure Ratio and the Compromise with Efficiency and Mass Flow 143
5.4 Stage Performance: Corrected Parameters 149
5.5 Multistage Compressors 151
5.6 Stall and Surge 154
5.7 Centrifugal Compressors 157
References 163
Problems 163

6
Turbines 166

6.1 Turbine Stage Characteristics 167
6.2 Turbine Blading 172
6.3 Turbine Cooling 174
6.4 Turbine Similarity 185
References 186
Problems 187

7
Structure of Engine Turbomachinery 189

7.1 Centrifugal Stresses 189
7.2 Gas Bending Loads on Blades 193
7.3 Thermal Stresses 195
7.4 Critical Speeds and Vibration 197
7.5 Blade Flutter 203
References 206
Problems 206

8
Component Matching and Engine Performance 208

8.1 Compressor-Turbine Matching: The Gas Generator 208
8.2 Gas Generator–Nozzle Matching 210
8.3 Engine-Inlet Matching and Distortion 211
8.4 Overall Performance 212
8.5 Control and Acceleration 213
References 220
Problems 220

9
Aircraft Engine Noise 222

9.1 Noise Sources: Unsteady Flow 224
9.2 Jet Noise 230
9.3 Turbomachinery Noise 237
9.4 Noise Measurement and Rules 245
References 249
Problems 250

10
Hypersonic Engines 251

10.1 Hypersonic Inlets 253
10.2 Heat Addition in High-Speed Flow 255
10.3 Heat Release Due to Chemical Reactions 257
10.4 Nozzle and Performance 262
10.5 Cooling 263
References 264
Problems 265

11
Propulsion Systems Analysis 266

11.1 Takeoff 266
11.2 Climb and Acceleration 267
11.3 Cruise 271
11.4 Maneuvering 273
References 273
Problems 273

Index 275

Preface

This book is intended to provide an introduction to the engineering of aircraft propulsion systems with the emphasis on the engine, rather than on the disciplines involved in its design. Because of the remarkable advances that have occurred since the large-scale introduction of gas turbine power plants into military aircraft in the 1950s and into commercial aircraft in the 1960s, a clear understanding of the characteristics of these devices is needed at the undergraduate or early graduate level. Such understanding is essential both for entrance to professional work in the industry and for graduate study in the field. The understanding of a sophisticated engineering system that involves the sciences of fluid mechanics, solid mechanics, chemistry, automatic control, and even psychology because of the problem of aircraft noise, also has intrinsic value apart from its practical applications. At present, the fundamental information required for such understanding is widely dispersed in the technical literature and subliterature. The aim of this book is to draw the information together in a unified form, so that the student can appreciate why aircraft propulsion systems have evolved to their present form and can thus be better prepared to contribute to their further evolution.

Automotive and stationary applications of gas turbines are growing rapidly. They use much the same technology as aircraft gas turbines; indeed they have benefited greatly from the aircraft engine developments of the last two decades. While this book is directed primarily at aircraft engines, the discussions of component technology are equally applicable to these other applications. Some treatment of cycles has also been included for automotive and stationary engines in chapters 2 and 3.

The approach in this book is to treat the propulsion system at successively increasing levels of sophistication, beginning with a phenomenological discussion in chapter 1 of the processes by which energy is converted from heat to mechanical energy to thrust.

Several types of engines are then discussed in chapter 2 in the framework of ideal cycle analysis, where the components of an actual engine are represented parametrically in the analysis without quantitative reference to the engine structure. Here the dependence of the engine's performance on compressor pressure ratio and turbine inlet temperature is established, as well as the trends of thrust and specific impulse with flight Mach number. The arguments are repeated more quantitatively in chapter 3 for a narrower spectrum of engines to convey the influence of nonidealities in the engine cycles.

Chapters 4, 5, 6, and 7 examine the mechanical characteristics required of each major engine component to achieve the parametric behavior assumed in the cycle analysis. At this step the enormous literature of the field must be abstracted and interpreted to clarify the important physical limitations and trends without submerging the reader in vast analyses and data correlations. Naturally, the presentation is strongly influenced by my own viewpoint. If it errs in detail or by omission at some points, I hope that the overview will help the serious student to correct these inadequacies for himself.

Chapter 8 synthesizes, from the component characteristics evolved in chapters 4, 5, and 6, a complete gas generator and a complete propulsion system. An attempt is made to treat in a reasonably uniform way the problems of engine control, inlet-engine and engine-nozzle matching, and inlet distortion, which so strongly dictate the ultimate performance of the system.

The mechanisms by which aircraft engines produce noise are discussed in chapter 9. At its present state of development, this subject is both highly mathematical and highly empirical. While the mathematics of this chapter is somewhat more advanced than that in other chapters, it should be understandable to a well-prepared college junior or senior. In any case, some care has been taken to make the physical arguments independent of the mathematics.

Since flight at very high Mach numbers leads to complex chemical behavior of the air as it passes through the engine, the possibilities for airbreathing propulsion at Mach numbers above 6 are discussed separately in chapter 10, where the thermochemistry of high-temperature combustion products is included.

Finally, chapter 11 deals with some of the simpler techniques of propulsion systems analysis, the tool used by the preliminary designer to determine which engine should be committed to the lengthy and costly process of development.

To understand this text, a student needs good first courses in gasdynamics, thermodynamics, and solid mechanics along with the appropriate mathematics. These subjects will not be reviewed here, but some of the results of compressible flow are collected at the beginning of chapter 4.

Though this book developed from a one-semester undergraduate course in aircraft engines at MIT, it contains more information than can reasonably be taught in one semester. A good one-semester undergraduate course in aircraft engines might consist of chapters 1 and 2 and the following selections from the remaining chapters:

Chapter 3: 3.1–3.5, 3.7
Chapter 4: 4.1, 4.2.2.1–4.2.2.3, 4.3, 4.4.3, 4.4.4
Chapter 5: 5.1, 5.2.2, 5.2.3, 5.2.5, 5.3, 5.4, 5.5, 5.6
Chapter 6: 6.1, 6.1.1, 6.2, 6.3, 6.3.1, 6.3.2, 6.4
Chapter 7: 7.1–7.3
Chapter 8: 8.1, 8.2, 8.4

The text in its entirety is suitable for first-year graduate students with no prior exposure to aircraft engines.

Acknowledgments

My understanding of aircraft propulsion systems has benefited from associations with many persons—from academe, from the industry, from NACA-NASA, and from the armed services—over the last two decades. Edward S. Taylor, James E. McCune, Jean F. Louis, Alojzy A. Mikolajczak, and Leroy H. Smith have been particularly helpful in formulating some of the arguments in chapter 5. Those familiar with the teaching and research of Frank E. Marble of the California Institute of Technology will recognize the powerful influence he has had, first as teacher and later as colleague, on me and on this book. Special thanks are due to my students at MIT, who provided the motivation for writing the book, and to my wife Vickie, who helped immeasurably in bringing it to completion.

Aircraft Engines and Gas Turbines

1
Introduction to Concepts

All aircraft engines and gas turbines are heat engines, in which thermal energy derived from the combustion of fuel with air (or derived, perhaps, from a nuclear reactor) is converted to useful work in one way or another. The efficiency of this conversion—the ratio of useful work output to energy input by the fuel or reactor—is of major and growing concern.

When the useful output of the gas turbine is in the form of shaft power used to drive a wheeled vehicle, a machine, or an electric generator, the efficiency may usually be characterized by the *thermal efficiency* familiar from thermodynamics, defined as the fraction of thermal energy input converted to mechanical work.

In aircraft propulsion the useful work of the engine is work done in propelling the aircraft. It is appropriate then to define a second efficiency, the propulsive efficiency, as the ratio of propulsive work to total mechanical work. Although analogous efficiencies of utilization can be defined for other applications of gas turbines, the propulsive efficiency is particularly important because it plays a dominant role in determining the configurations of aircraft engines. The different types of engines—ramjets, turbojets, turbofans, and turboprops—result from optimizing the overall efficiency, which is the product of thermal efficiency and propulsive efficiency, for different flight regimes.

Overall efficiency, however, is not the sole criterion for engine design. In aircraft engines weight and size are also important. Cost, while high for optimum engines, is important. Recently, takeoff noise has become a major problem for commercial aircraft operators, so that noise produced per unit of thrust has become an important criterion for engine design.

In automotive applications cost limits engines to much simpler and less efficient designs than those evolved for aircraft. For stationary applications reliability, efficiency, and cost are controlling, while size and weight are much less important.

The purpose of this chapter is to describe the fundamental characteristics of gas turbines that control and limit their adaptation to any of these applications. Some are thermodynamic, some fluid dynamic, some mechanical.

1.1 Thermal Efficiency

The conversion of thermal energy to mechanical energy is subject to the laws of thermodynamics. These laws determine an upper limit on the thermal efficiency that depends only on the temperatures at which heat is added to and rejected from the working fluid of the engine. Most gas turbines use the atmosphere as a heat sink, so that the minimum available heat rejection temperature is the atmospheric temperature, denoted by T_0. The maximum available heat addition temperature is in principle limited only by the characteristics of the combustion process (or nuclear reactor). In practice it may be limited by the temperature capabilities of materials. If this maximum heat addition temperature is denoted by T_m, the *maximum possible* thermal efficiency is that attained by a Carnot cycle operating between these temperature extremes, expressed by

$$\eta_c = 1 - \frac{T_0}{T_m}. \tag{1.1}$$

In the stratosphere (between 11 and 30 km altitude), T_0 is approximately 217°K. Current aircraft gas turbines have peak temperatures near 1500°K, so that η_c is approximately 0.85. Automotive and stationary gas turbines generally have peak temperatures below 1300°K, for reasons of cost and durability, and they reject heat at about 300°K, so for them η_c is about 0.77. Actual engines have thermal efficiencies lower than these. For comparison, the maximum possible efficiency for steam power plants is near 0.66.

1.2 Propulsive Efficiency

Unlike thermal efficiency the propulsive efficiency, representing conversion between two forms of mechanical energy, is limited only by the laws of mechanics and can in principle approach unity. It is defined as

$$\eta_p = \frac{\text{thrust power delivered to vehicle}}{\text{net mechanical power in exhaust}}.$$

The numerator is equal to the thrust multiplied by the flight velocity, while the denominator is the product of the mass flow and the increase in kinetic energy imparted by the engine to the airflow.

By the conservation of momentum, the force acting on the engine due to the flow through it is equal to the time rate of change of the momentum of the flow. If the mass flow per unit time is $\dot{m}$, the flight velocity is u_0, and the exhaust velocity is u_e, the thrust is

$$F = \dot{m}(u_e - u_0), \tag{1.2}$$

and the propulsive efficiency is

$$\eta_p = \frac{\dot{m}(u_e - u_0)u_0}{\dot{m}(u_e^2/2 - u_0^2/2)} = \frac{2u_0}{u_e + u_0}. \tag{1.3}$$

The propulsive efficiency decreases as the ratio of exhaust velocity to flight velocity increases. From (1.2), we can see that for a given mass flow and flight velocity, the thrust increases with u_e/u_0. Thus, a definite tradeoff must be made between propulsive efficiency and thrust per unit mass flow. This relationship, shown in figure 1.1, applies generally to all aircraft engines.

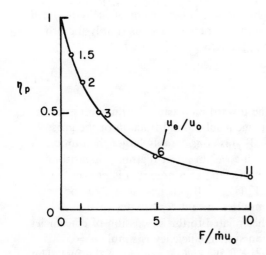

1.1 Propulsive efficiency as a function of thrust per unit of inlet air momentum, with ratio of exhaust velocity to flight velocity as parameter.

Increased mass flow in general implies increased engine size and weight, and it may also increase drag.

1.3 Specific Impulse and Range

The discussion of engine types in terms of efficiencies links cycle analysis and thermodynamics, thus providing an intuitive grasp of propulsion system characteristics; but the propulsion system efficiency is usually characterized in terms of the *specific impulse*, defined as the number of units of thrust produced per unit of fuel weight flow rate. This quantity enters directly into calculations of the fractional weight change of aircraft. It is denoted by I.

Suppose an aircraft is in steady, straight, and level flight. The thrust F must then equal the drag D. The aerodynamic performance of the airframe is characterized by its ratio of lift to drag L/D. Since the lift must equal the weight W of the aircraft, $F = W/(L/D)$. Now the weight of the aircraft decreases as fuel is consumed; the rate of decrease is $dW/dt = -F/I$, by the definition of I. Thus

$$\frac{dW}{dt} = -\frac{W}{I(L/D)},$$

and if I and (L/D) are constant in time, the flight duration t is given by

$$t = I(L/D) \log \frac{W_g}{W_g - W_f} \qquad (1.4)$$

where W_g is the initial (gross) weight and W_f is the weight of fuel consumed. It is usual to present this result in terms of range, which is simply the product of the flight duration and the flight velocity u_0, so that

$$\text{Range} = u_0 I(L/D) \log \frac{W_g}{W_g - W_f}. \tag{1.5}$$

Historically, much effort has gone toward increasing the range of aircraft. As a result, the fuel weight has become a substantial fraction of the gross weight, and the fraction $W_g/(W_g - W_f)$ is considerably larger than unity. In this case, structural weight or engine weight affect the range logarithmically, while I, u_0, and (L/D) affect it directly, so a premium is put on these factors. On the other hand, when $W_g/(W_g - W_f)$ is near unity, engine weight becomes as important as specific impulse, since it contributes to W_g.

The specific impulse can be further related to this discussion of efficiencies by noting that the overall propulsion system efficiency is simply $\eta = Fu_0/(-dW/dt)h = Fu_0/(F/I)h$, where h is the energy content of the fuel. Thus the factor $u_0 I$ in (1.5) is simply ηh, the product of the energy content of the fuel (in units such as ft-lb per lb or m-kg per kg) and the efficiency with which it is used. The value of h for liquid hydrocarbon fuels is about 4800 km. For hydrogen it is 14,300 km, and for methane, 5600 km.

1.4 Ramjets

Ramjets are conceptually the simplest of aircraft engines. Figure 1.2 is a schematic cross-sectional diagram of such an engine. Focusing for the present on the behavior of the airflow passing through the engine (indicated by the dashed inlet and exhaust streamtubes), we see that the schematic depicts only the internal functions of the engine. This engine consists of an inlet (diffuser), a combustor (burner), and a nozzle. The inlet decreases the flow velocity rel-

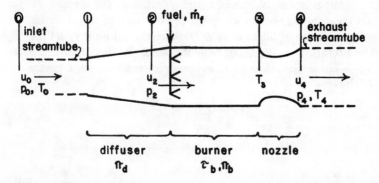

1.2 Schematic diagram of ramjet engine.

ative to the engine from the flight velocity u_0 to some smaller value u_2. The difference in kinetic energies of the air $(u_0^2/2 - u_2^2/2)$ per unit mass is converted to an increase in thermal energy, so that $T_2 > T_0$; at the same time, the pressure increases from p_0 to a higher value p_2. Fuel is then mixed with the air, and the mixture is burned in the combustor. If the velocity u_2 is small compared to the sonic velocity (the Mach number $M_2 \ll 1$), the combustion occurs at nearly constant pressure; the net result is that the thermal energy of the fluid increases, and its density decreases. In the nozzle the flow is expanded, ideally to the original pressure, with a consequent drop in temperature from T_3 to T_4 and an increase in kinetic energy $u_4^2/2 - u_3^2/2$. Since T_3 is larger than T_2, the difference in thermal energies between stations 3 and 4 is larger than that between stations 2 and 0; therefore, the change in kinetic energy in the nozzle is larger than that in the inlet, and u_4 is larger than u_0. The change in momentum $u_4 - u_0$ per unit mass flow provides the thrust.

The conversion of thermal energy to mechanical energy is represented ideally by a Brayton cycle, as shown in figure 1.3. This cycle may be thought of as a superposition of a number of Carnot cycles, indicated by the small rectangles, each with a temperature ratio $T_2/T_0 = T_3/T_4$. Accordingly the maximum possible efficiency of the cycle is

$$\eta_B = 1 - T_0/T_2. \tag{1.6}$$

The maximum efficiency can approach the limiting Carnot efficiency n_C only if T_2 approaches T_3, that is, if all temperature rise occurs in the inlet rather than in the combustor. The thermal efficiency of the ideal ramjet is thus con-

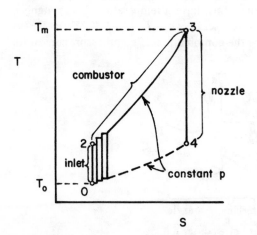

1.3 Temperature-entropy diagram of Brayton cycle for ramjet, with elementary Carnot cycles of which it is composed.

trolled by the inlet compression process, which governs the temperature ratio T_2/T_0. In the ideal case where $u_2 \ll u_0$ this ratio approaches the stagnation to static temperature ratio of the inlet flow, $T_2/T_0 = 1 + \frac{1}{2}(\gamma - 1)M_0^2$, so that

$$\eta_B = \frac{\frac{1}{2}(\gamma - 1)M_0^2}{1 + \frac{1}{2}(\gamma - 1)M_0^2}, \tag{1.7}$$

where M_0 is the flight Mach number and $\gamma = c_p/c_v$ is the ratio of specific heats. Thus for $M_0 < 1$, T_2/T_0 approaches unity and the thermal efficiency of the ramjet becomes small. It becomes a highly efficient engine for $M_0 > 3$.

In the ideal ramjet u_4/u_0 (and hence η_p) is determined by the combustor temperature ratio T_3/T_2. For a given u_0, increasing T_3 will increase thrust, but reduce propulsive efficiency.

1.5 Turbojets

The poor performance of the ramjet at low Mach number is improved in the turbojet engine, shown schematically in figure 1.4, by the addition of a compressor. The compressor raises the air pressure and temperature prior to combustion and thus improves the cycle efficiency. The ideal Brayton cycle for the turbojet is shown by the full lines in figure 1.5. The thermal efficiency is now given by

$$\eta_B = 1 - T_0/T_3. \tag{1.8}$$

If, for example, the compressor pressure ratio is 12, corresponding to an ideal compressor temperature ratio of 2.03, the ideal thermal efficiency is about 0.5. To drive the compressor the turbine must have a temperature drop roughly equal to the compressor temperature rise. Because $T_4 > T_3$, $T_4/T_5 < T_3/T_2$, and it follows that $p_5 > p_2$. Thus the combination of compressor, combustor,

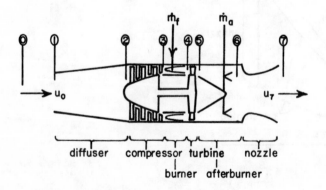

1.4 Schematic diagram of turbojet engine.

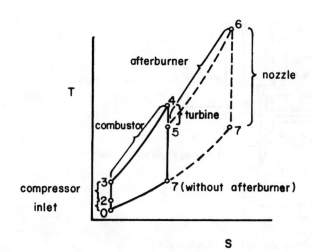

1.5   Temperature-entropy diagram for Brayton cycle of turbojet, with afterburning modification shown dashed.

and turbine, called a gas generator, produces a rise in pressure as well as in temperature of the airflow.

Materials limitations in the turbine at present restrict the turbine inlet temperature T_4 to values below those corresponding to a stoichiometric mixture of fuel and air in the combustor, so that the turbine exhaust gas contains considerable residual oxygen. Additional thrust can be obtained by adding fuel in an afterburner. The cycle for this modification is shown dashed in figure 1.5. Because this fuel is added at lower pressure (and temperature) than the fuel in the primary combustor, it is used less efficiently. The penalty at subsonic speeds is so large that afterburning is used only for short bursts of extra thrust. At Mach numbers of 2.5 or more, the afterburning turbojet becomes highly efficient because the pressure rise associated with diffusion in the inlet raises the nozzle pressure ratio to a high value, as in the ramjet.

The propulsive efficiency of a turbojet is determined in the same way as that of a ramjet, by the combustor temperature ratio. For an ideal engine η_p can be made to approach unity by letting T_4 approach T_3, but the thrust for a given engine size (or mass flow) becomes small, as indicated by figure 1.1, so that in practice this option is not applied.

1.6   Turbofans

A better way to improve the propulsive efficiency of the basic turbojet is offered by the turbofan, sketched in figure 1.6. Here, a second turbine is

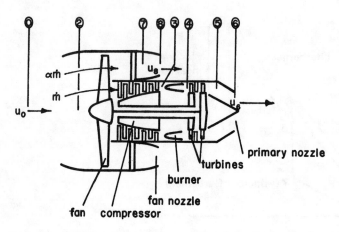

1.6 Schematic diagram of turbofan engine for subsonic flight.

added downstream of the compressor-drive turbine, and the power is used to drive a fan that pumps air through a secondary nozzle. By this means a portion of the energy of the primary jet is removed, its velocity is reduced, and the energy is transferred to the fan airstream. Thus the effective value of $F/\dot{m}u_0$ is reduced and η_p is increased as in figure 1.1; the penalty in engine weight is less than would be caused by decreasing T_4. The turbofan powers most modern subsonic transport aircraft; it has replaced both the turbojet and turboprop in this application. Modified to include afterburning in the duct airflow or in the mixed primary and fan flows, it also powers many high-performance military aircraft and is a major contender for the supersonic transport.

1.7 Shaft Engines: Regeneration

For low-speed flight vehicles, where prohibitive size and weight of duct would be required to yield a good propulsive efficiency with a turbofan, the turboshaft engine is used. Here, most of the useful work is extracted from the exhaust gas by a turbine; it often rotates on a separate shaft from the gas generator, as sketched in figure 1.7, and may drive a propeller, helicopter rotor, the wheels of a truck, or any other machine.

In land or marine applications, where weight and size are not primary considerations, *regeneration* is used to increase the thermal efficiency beyond that attainable with a simple Brayton cycle. The regenerator is a heat exchanger that transfers heat from the exhaust gas to the compressor discharge air. As sketched in figure 1.8, this transfer can be accomplished with a rotating heat storage matrix. The regenerative Brayton cycle is shown in figure

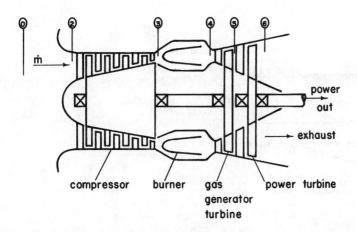

1.7   Schematic diagram of shaft turbine.

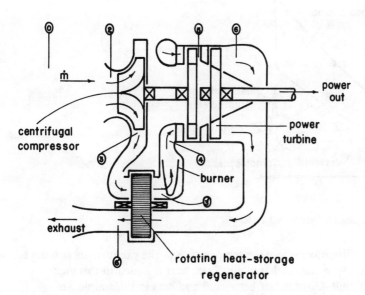

1.8   Shaft turbine with centrifugal compressor, and rotating matrix regenerator.

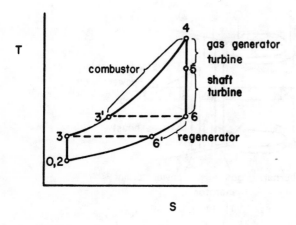

1.9   Temperature-entropy diagram for regenerative Brayton cycle.

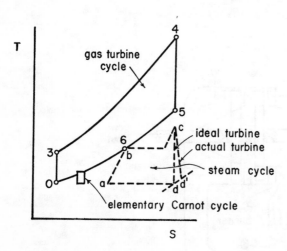

1.10   A gas turbine-steam combined cycle.

1.9 for the ideal situation where the temperature of the exhaust gas is reduced to that of the compressor discharge air. Since heat is added in this engine only between 3′ and 4, instead of between 3 and 4 as in the simple gas turbine, the regenerated engine will have higher efficiency than the simple one for the same compressor pressure ratio. In other words, for a fixed turbine inlet temperature T_4, an acceptable efficiency can be obtained with a lower compression ratio in the regenerated than in the nonregenerated engine. Since efficient, high pressure ratio compressors are complex and expensive, regenerators are used in engines for automotive applications where cost is a major consideration.

1.8 Stationary Gas Turbines: Topping

The turbine inlet temperature of modern gas turbines is considerably higher than the peak steam temperature in steam power plants. Depending upon the compression ratio of the gas turbine, the turbine exhaust temperature may be high enough to permit efficient generation of steam using the waste heat from the gas turbine. Such an arrangement is referred to as a gas turbine-combined cycle power plant. The cycle is shown in figure 1.10. It is capable of very high efficiencies when the turbine inlet temperature of the gas turbine is high. It's advantage over the regenerative gas turbine is that the steam boiler is easier to manufacture and maintain than the regenerator.

1.9 Energy Exchange, Mach Number, Reynolds Number

Four types of energy exchange have been implicitly involved in the above descriptions of engines. These are (1) the exchange within a flowing fluid of kinetic energy for thermal energy or vice versa; (2) transfer of energy to or from a fluid by forces acting on moving blades; (3) the conversion of chemical energy to thermal energy; and (4) the transfer of thermal energy from solid bodies to flowing fluids.

The exchange from kinetic energy to thermal energy occurs when the momentum of a fluid is changed by pressure forces. The increasing pressure compresses the gas and the compression work appears as an increase in internal (thermal) energy, according to the first law of thermodynamics. The Mach number is defined as the ratio of the flow velocity to the velocity of sound in the fluid $M = u/a$. When squared, it may be viewed as a measure of the ratio of kinetic energy to thermal energy of the fluid. Thus

$$\frac{1}{2}(\gamma - 1)M^2 = \frac{u^2/2}{c_p T}. \tag{1.9}$$

It follows that if process (1) is to be important, changes in $\frac{1}{2}(\gamma - 1)M^2$ that are large compared to unity must occur. The ramjet depends entirely on this process of energy exchange, and this is the reason it must operate at Mach numbers above unity.

The second process appears in the turbojet, the turbofan, the turboprop, and all other devices using fluid dynamic machinery. The air flow over a blade in a compressor, for example, exerts a force on the blade. If the blade moves in a direction opposite the force, then the blade does work on the air, increasing its mechanical energy. Process (1) may take place at the same time, so that the overall change in fluid energy appears partly as kinetic energy and partly as thermal energy. Now the force exerted on a body per unit area by a fluid is proportional to $\rho u^2/2$, where ρ is the fluid density and u is the velocity, which may be taken to be the same order as the velocity of the body. The power delivered to the fluid by the body, per unit area, is then of the order of $\rho u^3/2$. Thermal energy of the fluid is convected by the body at the rate $\rho u c_p T$ per unit area. Thus the ratio of energy addition by the body to convected thermal energy per unit time and area is

$$\frac{\rho u^3/2}{\rho u c_p T} = \frac{u^2}{2c_p T} = \tfrac{1}{2}(\gamma - 1)M^2, \tag{1.10}$$

and it can be seen that the Mach number plays the same key role in process (2) as in process (1). For the moving blades of the compressor or turbine to effectively exchange energy with the air, they should move at a Mach number of unity or more.

Process (3) is so familiar that it requires no elaboration, but process (4) requires some discussion. In gas turbines we are concerned primarily with convective heat transfer, that is, heat transfer that occurs between a solid surface and a fluid because of the motion of the fluid over the surface. The thermal effects of the surface on the fluid, like the viscous effects, are confined to a region near the surface that is thin compared to the characteristic length of the surface when the Reynolds number is large. That is, if we consider the flow over a flat plate of length L, as sketched in figure 1.11, with fluid density ρ, velocity u, and viscosity μ, then for laminar flow the viscous effects penetrate a distance δ_μ of order

$$\frac{\delta_\mu}{L} \sim \left(\frac{\mu}{\rho u L}\right)^{1/2} = \left(\frac{1}{Re}\right)^{1/2}.$$

If the fluid has a Prandtl number $c_p\mu/k$ near unity, where k is the thermal conductivity, or if the flow is turbulent, the thermal effect of the plate penetrates a distance $\delta_k \approx \delta_\mu$. In most of the components of a gas turbine, we wish to minimize viscous effects; hence we desire large Re and thin boundary layers. But in a regenerator the thermal effect must penetrate the entire flow, so either Re must be small or the ratio of spacing between heat transfer

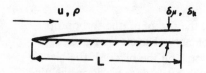

1.11 The penetration of viscous and thermal effects into a flowing fluid.

surfaces to their flow length must be small, of order $1/Re$. In either case the result tends to be a bulky and heavy device compared to the compressor and turbine. For this reason regenerators are not found in aircraft engines.

1.10 Stresses

Since the speed of sound in air is about 340 m/sec at normal conditions, the blading of compressors and turbines should have velocities near 340 m/sec or more. This requirement has forced the designer of gas turbine engines to cope with materials, vibration, and stress problems of a very high order. By contrast, the piston speed of a typical "high-speed" gasoline engine is only about 15 m/sec.

Some appreciation for the problem can be had by considering a prismatic bar rotating about an axis at one of its ends, as in figure 1.12, with an angular velocity ω. The stress in the bar at any radius, due to centrifugal forces, will be

$$\sigma = \int_r^{r_T} \rho\omega^2 r \, dr = (\rho\omega^2/2)[r_T^2 - r^2].$$

For $r \ll r_T$, the stress is

$$\sigma/\rho = (\omega r_T)^2/2. \tag{1.11}$$

For ωr_T to be 340 m/sec, it is necessary to have a material with the ratio of properties σ/ρ of the order of 6×10^4 m²/sec². For steel, with a density of 8000 kg/m³, this implies a stress of $\sigma = 4.8 \times 10^8$ N/m², close enough to the potential limit of the material that great sophistication and care in design are required. The problem is compounded in the turbine by the exposure of the rapidly rotating turbine blades to hot exhaust gases. This factor, probably more than any other, has limited the performance of aircraft gas turbines.

1.11 Noise

Acoustical noise is radiated from regions of fluctuating pressure, which may be produced in many ways. There are at least four sources of strong unsteady

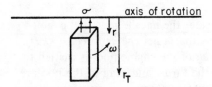

1.12 Bar, rotating about axis through its end, illustrating stress in rotating parts of engines.

flows in aircraft engines. These are the motion of bladed rotors (fans in particular), the passage of moving compressor and turbine blades past neighboring stationary blades, combustion (which results in local expansions of the burning gases), and turbulent mixing of high-velocity gases (such as the mixing of propulsive jets with the ambient air). The last, which leads to jet noise, is a direct manifestation of propulsive inefficiency, since it represents the energy radiated from the jet as it dissipates its excess kinetic energy by turbulent mixing with the air. As the bypass ratio of turbofans has been increased, largely to improve η_p, their jet noise has, happily, decreased. Noise from turbomachinery has its origins in part in the very blade forces that lead to energy exchange with the fluid. The unsteadiness that must be present if these forces are to produce noise is due in some cases simply to the rotation of the compressor and in other cases to the passage of moving blades past nearby stationary objects or through their wakes. The "buzz saw" noise of the B 747 on takeoff is in the former category, while the high-pitched whine more usually associated with turbojet engines is in the latter.

Much engine noise results from the pressure fluctuations associated with turbulence in wakes, boundary layers, and a multitude of unidentified other places.

1.12 Thrust and Drag

Conventionally, the forces acting on an aircraft in its direction of motion are divided into two parts, thrust and drag. The thrust is defined as the part of the force resulting from changes in the momentum (or pressure) of the air that *flows through the engine*. The drag is the force resulting from changes in the momentum of the air that *flows over the exterior of the vehicle*. In some cases this distinction is ambiguous, but in general it is useful and indeed essential to avoid confusion.

The definitions of thrust and drag do not imply that the drag is independent of the engine's operation or that the thrust is not influenced by the flow over the exterior of the aircraft. Especially in supersonic aircraft, the interaction of the internal (engine) and external airflows must be accounted for in determining either thrust or drag, and when the engines are embedded in the wing roots or fuselage the thrust and drag accounting requires an understanding of the flow over the entire aircraft.

To simplify the problem somewhat consider the nacelle mounted engine shown schematically in figure 1.13. Assume that a net *engine force*, $F - D_e$, which represents the sum of thrust and drag of the engine pod, is carried to the aircraft structure by the strut, so that the thrust and drag of the pod are defined independently of the aircraft. A control volume is constructed, closed at the front and sides by surfaces sufficiently far from the engine pod so that

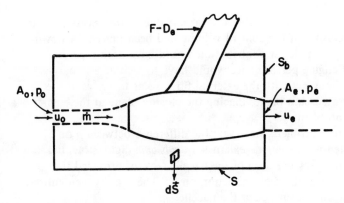

1.13 Schematic of pod-mounted engine, illustrating conventions for separating thrust (F) and drag (D_e) and method for computing thrust.

the pressure and flow velocity have their free-stream values p_0 and u_0. The volume is closed at the rear by a plane through the exit nozzle of the engine. The force $F - D_e$ must equal the time rate of change of the momentum of the contents of the control volume. If the flow is steady, this balance may be expressed

$$F - D_e - A_e(p_e - p_0) - \int_{S_b} (p - p_0)\, dS = \int_S \rho u \mathbf{u} \cdot d\mathbf{S}$$

where $\mathbf{u}$ is the flow velocity, u is its component in the flight direction, and $d\mathbf{S}$ is the differential element of the surface S. Because the aft plane of the control volume is close to the engine, the pressure on the aft plane is not equal to the free-stream pressure. The pressure of the *internal* (engine) flow crossing this plane has been denoted p_e, for exit. In accordance with the definitions of thrust and drag, this balance is divided into two parts,

$$F - A_e(p_e - p_0) = \int_{A_e} \rho u \mathbf{u} \cdot d\mathbf{S} + \int_{A_0} \rho u \mathbf{u} \cdot d\mathbf{S},$$

$$D_e + \int_{S_b} (p - p_0)\, dS = -\int_{S - A_e - A_0} \rho u \mathbf{u} \cdot d\mathbf{S}.$$

If ρ and u are uniform over the exit plane of the engine nozzle, then ρu in the first integral of the top equation is simply the engine exit mass flux and u is the exit velocity. Similarly, in the second integral ρu is the free-stream mass flux, so that the thrust equation is

$$F = \dot{m}_e u_e - \dot{m}_0 u_0 + A_e(p_e - p_0). \tag{1.12}$$

This result will form the basis for estimates of engine thrust throughout the following chapters. It indicates that the thrust is composed of a part due to the excess of momentum in the exhaust plus a part due to the excess pressure. If the downstream closure of the control volume had been placed far downstream, where $p_e \to p_0$, the latter contribution would have been zero, but then the analysis of engine performance would have to include an analysis of the mixing of the exhaust jet with the external flow, so as to arrive at the velocity over the downstream plane. Placing the closure plane at the engine exit eliminates this problem.

It introduces another problem, however; the difference between p on S_b and p_0 can be affected by the engine exhaust. In *subsonic* flight, potential flow theory tells us that the pressure drag of the nacelle is zero, provided the external flow is parallel to the flight direction on S_b. The drag is then entirely due to viscous shear on the surface of the nacelle.

In supersonic flow the presence of shock waves in the external flow leads to an entropy rise, which appears as either a pressure defect on the plane S_b or as a velocity defect there. Either will lead to an increase in drag. To the extent that deflections of the external airflow are caused by the engine airflow, the drag may be thought of as due to the engine rather than to the airframe. At times both airframe designers and engine designers have been loath to accept responsibility for this interface.

Spillage drag at the inlet and base drag at the exit are examples of such interaction problems. Figure 1.14 shows the flows that result in the excess drag. If the engine cannot accept all the flow the inlet would normally capture, a shock forms that aids in turning the flow around the outside of the diffuser; but in the process it increases the entropy of the air, thus creating a drag in the external flow. If the nozzle area of the engine must be reduced to a value smaller than the base area of the nacelle for the engine to operate properly, the external flow must turn through an expansion and

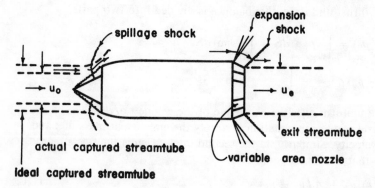

1.14 The flow over an engine at supersonic speed, showing origins of inlet spillage drag and base drag.

shock system to fill the space not occupied by the jet. The shocks result in entropy production and drag.

1.13 Some Engines in Cutaway

These engine types and the evolutionary trends will be discussed in much greater detail in the remainder of this book. At this point and later, it will be helpful to refer to the geometry of some actual engines. Fortunately, the engine manufacturers have developed elegant cutaway drawings of each of their engines, and a few of these are presented in figures 1.15–1.19.

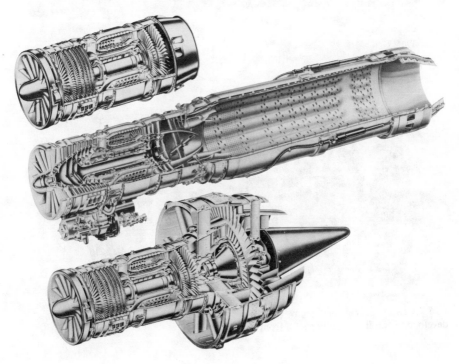

1.15 A small, single-shaft gas generator (GE J85) and the afterburning turbojet and aft-fan turbofan based on it.

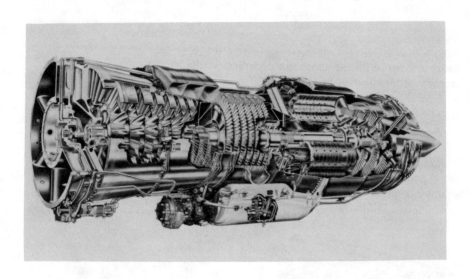

1.16   A two-spool turbojet engine (P & W JT3) and the turbofan (P & W JT3D) developed from it by addition of a front fan.

1.17 A modern, afterburning, low-bypass ($\alpha \approx 1$) turbofan engine (P & W F100 PW–100).

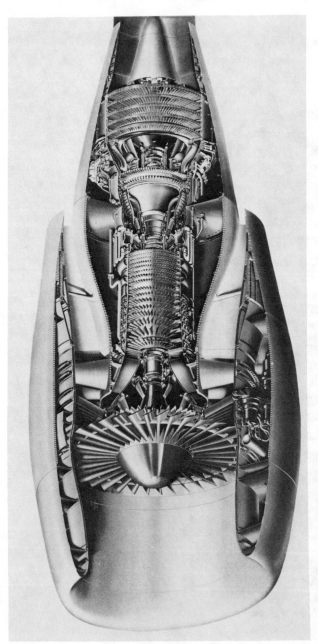

1.18 A modern, high-bypass ($\alpha \approx 5$) turbofan engine (GE CF6–6) in nacelle installation.

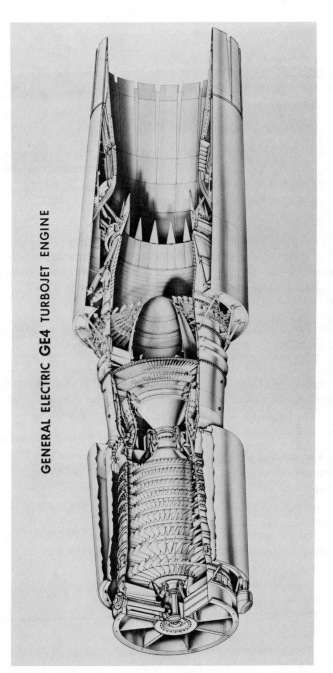

GENERAL ELECTRIC **GE4** TURBOJET ENGINE

1.19 A large, afterburning turbojet (GE4) designed and built for the Boeing 2707 SST, but not produced in quantity.

Problems

1.1 Assuming constant specific heats c_p and c_v, derive expressions for the thermal efficiency of the ideal ramjet and for the net work per unit of air flow. By equating the latter to the increase in jet kinetic energy, show that the thrust per unit of air mass flow is

$$F/\dot{m} = u_0\{1 + (2c_pT_0/u_0^2)[(T_2/T_0) - 1][(T_3/T_2) - 1)]\}^{1/2}.$$

1.2 Derive an expression for the specific impulse of an aircraft engine operating on a Carnot cycle, as a function of the peak-to-ambient temperature ratio, thrust per unit of mass flow, and flight velocity. Plot I as a function of M_0 with $F/\dot{m}a_0$ (where a_0 is the speed of sound) as a parameter for $T_m = 1500°K$, $T_0 = 217°K$, $h = 4.30 \times 10^7$ Joule/kg.

1.3 Write expressions for the engine pressure ratio p_5/p_2 and engine temperature ratio T_5/T_2 of a turbojet in terms of the compressor temperature ratio T_3/T_2 and turbine inlet temperature ratio T_4/T_2. Assume the compression and expansion are reversible and adiabatic. Compute p_5/p_2 and T_5/T_2 for $T_3/T_2 = 2$, $T_4/T_2 = 5$.

1.4 Compute the maximum Mach number of the piston motion, based on the speed of sound in ambient air, for a piston engine with a stroke of 0.1 m, rotating at 4000 rpm. Compare to (1.10).

1.5 Following the argument of (1.10), compute the stress in a thin ring rotating with angular speed ω about its axis of symmetry. If the ultimate strength is 1.36×10^9 Nm^{-2} and the density is 8×10^3 kg-m^{-3}, what is the limiting peripheral velocity?

1.6 The bypass ratio α as defined in figure 1.6 is a key design parameter for aircraft engines. Using the definitions of thermal and propulsive efficiencies, show that it can be expressed as

$$\alpha = (q/c_pT_0)[1/(\gamma - 1)M_0^2]\eta(\dot{m}_0u_0/F) - 1$$

where q is the heat added per unit mass of airflow through the core engine and η is the overall efficiency. Assume that the core and fanjet velocities are equal and that all fan work is converted to fanjet kinetic energy.

2
Ideal Cycle Analysis: Trends

Cycle analysis is the study of the thermodynamic behavior of the air as it flows through the engine without regard for the mechanical means used to effect its motion. Rather than dealing with the inlet, compressor, and turbine themselves, we characterize them by the results they produce. Thus, for example, the compressor will be specified by a stagnation pressure ratio and efficiency. The behavior of a real engine is in fact determined by its geometry, so that for a given geometry the compressor will produce some well defined pressure ratio, depending of course on other operating parameters such as speed and airflow. Thus when we plot curves of thrust and specific impulse versus, say, Mach number, we are not portraying the behavior of a real engine. Rather, each point on such a curve represents a different engine. We say the cycle analysis represents a "rubber engine."

The main purpose of cycle analysis is to determine which characteristics to choose for the various engine components to best satisfy a particular need. The value of cycle analysis therefore depends directly on the realism with which the components are characterized. If a compressor is specified by pressure ratio and efficiency, for example, and the analysis purports to select the best pressure ratio for a particular mission, the choice may depend on the variation of efficiency with pressure ratio. Unless a realistic variation of efficiency with pressure ratio is included in the analysis, its conclusion will be worthless.

We shall develop the cycle analysis in stages; we begin with the simplest possible set of assumptions, where all components are taken to be ideal, and recognize that only the simplest of conclusions can be deduced. Our purpose is to portray the characteristics of the several types of aircraft engines and gas turbines in the simplest possible way so that they can be compared. Following this, realistic assumptions as to component losses will be introduced in chapter 3 and the analysis repeated, to demonstrate the methods for choosing loss-sensitive parameters.

Thus, we assume
1. that compression and expansion processes in inlet, compressor, turbine, and nozzle are *isentropic*;
2. that combustion occurs at constant pressure;
3. that the working fluid is a perfect gas, with constant specific heat; and
4. that the exhaust nozzles expand the engine exhausts completely to ambient pressure.

It will not be possible to treat all possible engine cycles here; neither the author's nor the reader's patience would permit it. The hope is that the methods exemplified will allow the reader to carry out analyses according to his own need.

2.1 Stagnation Temperature and Pressure

In this book continual use will be made of the concepts of stagnation pressure and temperature. The reader is referred to the texts on gas dynamics for extensive discussions of these quantities. It will suffice here to define the stagnation temperature as the temperature reached when a steadily flowing fluid is stagnated (brought to rest) adiabatically. If T_t denotes the stagnation temperature, T the static (thermodynamic) temperature, and u the flow velocity, it follows from the conservation of energy that

$$c_p T_t = c_p T + u^2/2.$$

Introducing the Mach number $M = u/(\gamma RT)^{1/2}$, where R is the gas constant, we get

$$T_t = T(1 + \tfrac{1}{2}(\gamma - 1)M^2). \tag{2.1}$$

The stagnation pressure p_t is defined as the pressure reached if the stream is brought to rest isentropically as well as adiabatically. Then since $p_t/p = (T_t/T)^{\gamma/(\gamma-1)}$,

$$p_t/p = (1 + \tfrac{1}{2}(\gamma - 1)M^2)^{\gamma/(\gamma-1)}. \tag{2.2}$$

As ratios of stagnation temperatures and pressures will be used extensively, a special notation will be adopted for them. We denote a ratio of stagnation pressures across a component of the engine by π, with a subscript indicating the component; d for diffuser, c for compressor, b for burner, t for turbine, n for nozzle, f for fan. Similarly τ will denote a ratio of stagnation temperatures.

Stagnation temperatures divided by ambient static temperature will be denoted θ with a subscript. Thus, $T_{t0}/T_0 = 1 + \tfrac{1}{2}(\gamma - 1)M_0^2$ will be θ_0, while $T_{t4}/T_0 = \theta_t$ (for turbine inlet).

Stagnation pressures divided by ambient static pressure will be denoted δ, so that $p_{t0}/p_0 = (1 + \tfrac{1}{2}(\gamma - 1)M_0^2)^{\gamma/(\gamma-1)}$ will be δ_0, and we note that $\delta_0 = \theta_0^{\gamma/(\gamma-1)}$.

2.2 The Ramjet

With the notation and station numbers of figure 1.2, the thrust of the ideal ramjet is given by

$$F/\dot{m}u_0 = u_4/u_0 - 1,$$

and the specific impulse is

$$I = F/g\dot{m}_f$$

if $\dot{m}_f$ is the fuel mass flow rate and g is the acceleration of gravity. To compute u_4/u_0, we note first that the nozzle exit stagnation temperature, T_{t4}, is given by

$$T_{t4} = T_4(1 + \tfrac{1}{2}(\gamma - 1)M_4^2) = T_0\theta_0\tau_b.$$

Similarly, the exit stagnation pressure is

$$p_{t4} = p_4(1 + \tfrac{1}{2}(\gamma - 1)M_4^2)^{\gamma/(\gamma - 1)} = p_0\delta_0.$$

Since the nozzle is ideally expanded, $p_4 = p_0$, so from this last relation,

$$(1 + \tfrac{1}{2}(\gamma - 1)M_4^2)^{\gamma/(\gamma - 1)} = \delta_0 = (1 + \tfrac{1}{2}(\gamma - 1)M_0^2)^{\gamma/(\gamma - 1)}$$

and $M_4 = M_0$. It follows that $T_4/T_0 = \tau_b$. Now $u_4/u_0 = (M_4/M_0)(T_4/T_0)^{1/2} = \sqrt{\tau_b}$ so that finally

$$F/\dot{m}a_0 = (\sqrt{\tau_b} - 1)M_0. \tag{2.3}$$

Define h as the heating value of the fuel; then from an energy balance across the burner, $\dot{m}c_pT_0\theta_0 + \dot{m}_fh = (\dot{m} + \dot{m}_f)c_pT_0\theta_0\tau_b$ and if the fuel/air ratio $f = \dot{m}_f/\dot{m} \ll 1$, then $f = (c_pT_0\theta_0/h)(\tau_b - 1)$ and the specific impulse is

$$I = \left(\frac{a_0h}{gc_pT_0}\right)M_0\frac{\sqrt{\tau_b} - 1}{\theta_0(\tau_b - 1)}. \tag{2.4}$$

Equations (2.3) and (2.4) exhibit the dependence of F and I on the Mach number and burner temperature ratio, showing that for given τ_b, both F and I increase linearly with M_0 for M_0 small. For a given M_0, F increases monotonically with τ_b, while I decreases from a limiting value of (a_0h/gc_pT_0) $(M_0/2\theta_0)$ attained for $\tau_b \to 1$.

A case of special interest is that where f has its stoichiometric value. For hydrocarbon fuels, T_{t3} is then of the order 2500°K for low M_0. Increasing M_0 increases θ_0, but T_{t3} does not increase much because the combustion products tend to dissociate. The result is that $\theta_b = T_{t3}/T_0$ is sensibly independent of M_0, and we can find the dependence of F and I on M_0 by putting $\tau_b = \theta_b/\theta_0$ where θ_b is held constant in (2.3) and in the numerator of (2.4). But the $\tau_b - 1$ in the denominator of (2.4) represents the fuel/air ratio, so it must be replaced by $(\tau_b - 1)_{\text{stoich}} = f_{\text{stoich}}h/c_pT_0\theta_0$. The result is that for the stoichiometric ramjet we have

$$F/\dot{m}a_0 \approx (\sqrt{(\theta_b/\theta_0)} - 1)M_0,$$

$$I \approx (a_0/gf_{\text{stoich}})(\sqrt{(\theta_b/\theta_0)} - 1)M_0,$$

and I is simply a constant times $F/\dot{m}a_0$, which is shown in figure 2.1 for $\theta_b = 10$. We see that the thrust per unit mass flow and specific impulse peak at about $M_0 = 2.5$, so that the best operating range for hydrocarbon fueled ramjets is between $M_0 = 2$ and $M_0 = 4$.

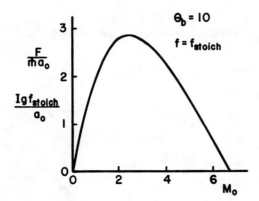

2.1 Thrust per unit mass flow and specific impulse for simple ramjet with stoichiometric combustion.

In the stratosphere, $a_0 \approx 296$ m/sec, so that $F/\dot{m}$, in units of kilograms of thrust per kilogram per second of airflow, is about 83 seconds at $M_0 = 2.5$. For stoichiometric combustion $f \approx 0.067$, so the group $a_0/gf_{\text{stoich}} \approx 452$, and the peak I is about 1310 seconds.

The significance of $F/\dot{m}a_0$ can be better understood if we realize that the mass flow which an engine of given size will ingest is a strong function of Mach number and altitude. The density of the atmosphere varies roughly exponentially with altitude; a useful approximation is

$$\frac{\rho_0}{\rho_0(h = 0)} = \exp(-h/9144),$$

where h is measured in meters. If an engine captured a streamtube (figure 1.2) of a constant area A_0 as M_0 varied then the mass flow would be simply $\dot{m} = \rho_0 u_0 A_0$. But it is more nearly true that the Mach number of the flow inside the engine at some point is constant. In the ramjet this point might be the combustor entrance; in a turbojet, the compressor inlet. If this is the case, the mass flow will depend on the Mach number at that internal point, the flow area, and the stagnation pressure and temperature. The ratio of ρu to $(\rho u)^*$, the value for $M = 1$, is a function only of M, the familiar A^*/A of channel flow theory. At standard sea level conditions, $(\rho u)^* \approx 239.2$ kg/m^{2}sec, and $(\rho u)^*$ is proportional to the stagnation pressure and to the negative square root of the stagnation temperature. Thus

$$\rho u/(\rho u)^*_{h=0} = M \left[\frac{\frac{1}{2}(\gamma + 1)}{1 + \frac{1}{2}(\gamma - 1)M^2} \right]^{(\gamma+1)/2(\gamma-1)} [1 + \frac{1}{2}(\gamma - 1)M_0^2]^{(\gamma+1)/2(\gamma-1)}$$
$$\cdot \exp(-h/9144)$$
$$\equiv \mu(M, M_0) \exp(-h/9144).$$

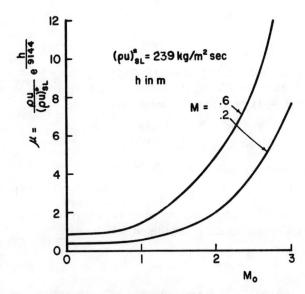

2.2   Engine mass flow density divided by choked flow density at sea level static conditions, as a function of flight Mach number M_0 and Mach number in engine M. h is altitude in meters.

The variation of this mass flow capability with M_0 is shown for two values of M in figure 2.2 for M_0 up to 3. Above this value, nonideal behavior of the diffuser completely destroys the validity of the trend.

   Multiplying the thrust curve of figure 2.1 by this mass flow factor, we see that the thrust of a ramjet actually should rise very steeply up to Mach numbers well above 3, for a given altitude.

2.3   The Turbojet

If we define π_c and τ_c as the stagnation pressure and temperature ratios across the compressor, that is, $\pi_c = p_{t3}/p_{t2}$ and $\tau_c = T_{t3}/T_{t2}$, and similarly, π_t and τ_t as the ratios across the turbine, we may proceed to analyze the turbojet as we did the ramjet. From the station numbers of figure 1.4, the nozzle outlet stagnation temperature is

$$T_{t7} = T_7\{1 + \tfrac{1}{2}(\gamma - 1)M_7^2\} = T_0\theta_0\tau_c\tau_b\tau_t$$

where the nozzle has been assumed adiabatic, so $\tau_n = 1$. Similarly,

$$p_{t7} = p_7\{1 + \tfrac{1}{2}(\gamma - 1)M_7^2\}^{\gamma/(\gamma - 1)} = p_0\delta_0\pi_c\pi_t.$$

Since the nozzle is ideally expanded, $p_7 = p_0$; thus

$$1 + \tfrac{1}{2}(\gamma - 1)M_7^2 = (\delta_0\pi_c\pi_t)^{(\gamma - 1)/\gamma} \tag{2.5}$$

and

$$T_7/T_0 = (\theta_0 \tau_c \tau_b \tau_t)/(\delta_0 \pi_c \pi_t)^{(\gamma-1)/\gamma}.$$

Since the flows through compressor and turbine are isentropic, $\tau_c = \pi_c^{(\gamma-1)/\gamma}$ and $\tau_t = \pi_t^{(\gamma-1)/\gamma}$. Furthermore, $\theta_0 = \delta_0^{(\gamma-1)/\gamma}$, so we find $T_7/T_0 = \tau_b$. From (2.5), $M_7^2 = [2/(\gamma-1)][\theta_0 \tau_c \tau_t - 1]$, and from the definition of θ_0, $M_0^2 = [2/(\gamma-1)][\theta_0 - 1]$,

$$u_7/u_0 = (M_7/M_0)(T_7/T_0)^{1/2} = [\tau_b(\theta_0 \tau_c \tau_t - 1)/(\theta_0 - 1)]^{1/2}.$$

The thrust of the turbojet is then given by $F = \dot{m} u_0(u_7/u_0 - 1)$ or

$$F/\dot{m}a_0 = M_0\{[\tau_b(\theta_0 \tau_c \tau_t - 1)/(\theta_0 - 1)]^{1/2} - 1\}. \tag{2.6}$$

This expression is not yet complete, as we must recognize that the power of the turbine equals that of the compressor. This condition, which relates τ_t to τ_c, can be written

$$\dot{m} c_p(T_{t3} - T_{t2}) = \dot{m} c_p(T_{t4} - T_{t5}).$$

Because the absolute magnitude of T_{t4} is generally limited by material temperature-stress capabilities or by cooling technology, it is useful to define a dimensionless temperature that represents this limitation. Let $\theta_t = T_{t4}/T_0$; then the compressor-turbine power balance can be solved for τ_t:

$$\tau_t = 1 - (\theta_0/\theta_t)(\tau_c - 1).$$

Substituting this expression in (2.6) and rearranging somewhat gives

$$F/\dot{m}a_0 = M_0(\{[\theta_0/(\theta_0 - 1)][(\theta_t/\theta_0 \tau_c) - 1](\tau_c - 1) + \theta_t/(\theta_0 \tau_c)\}^{1/2} - 1), \tag{2.7}$$

which gives the thrust per unit of mass flow as a function only of M_0, τ_c, and θ_t.

To obtain an expression for the specific impulse, an energy balance across the combustion chamber is needed:

$$\dot{m} c_p(T_{t4} - T_{t3}) = \dot{m}_f h,$$

where, as for the ramjet, $\dot{m}_f$ is the fuel mass flow and h is its energy content. Thus, $\dot{m}_f/\dot{m} = (c_p T_0/h)(\theta_t - \theta_0 \tau_c)$, and

$$I = F/g\dot{m}_f$$

$$= \frac{(a_0 h/g c_p T_0)M_0\{[\theta_0/(\theta_0 - 1)][(\theta_t/\theta_0 \tau_c) - 1](\tau_c - 1) + \theta_t/(\theta_0 \tau_c)\}^{1/2} - 1}{\theta_t - \theta_0 \tau_c} \tag{2.8}$$

so that I depends on the fuel and air properties, through the first (dimensional) factor, and otherwise only on M_0, τ_c, and θ_0 as did $F/\dot{m}a_0$.

A good deal can be learned from study of (2.7) and (2.8). First, for $M_0 \to 0$ the form of $F/\dot{m}a_0$ is indeterminate since $\theta_0 \to 1$. But putting $\theta_0 = 1 + \frac{1}{2}(\gamma - 1)M_0^2$ and expanding in M_0^2, we readily find

$$F/\dot{m}a_0 = \{[2/(\gamma - 1)][(\theta_t/\tau_c) - 1](\tau_c - 1)\}^{1/2} \qquad (M_0 = 0) \qquad (2.9)$$

and

$$I = (a_0 h/g c_p T_0) \frac{\{[2/(\gamma - 1)][(\theta_t/\tau_c) - 1](\tau_c - 1)\}^{1/2}}{\theta_t - \tau_c} \qquad (M_0 = 0). \qquad (2.10)$$

Both the thrust and impulse are nonzero at $M_0 = 0$, in contrast to the ramjet. Note that for $\tau_c \to 1$, both F and I approach zero, as this is the limit of the ramjet. A little study of (2.7) and (2.8) will show that they revert to (2.3) and (2.4) when $\tau_c \to 1$.

From the first two factors involving τ_c in (2.7), it is clear that for given M_0 and θ_t there is a value of τ_c that maximizes $F/\dot{m}a_0$, since the first factor decreases with increasing τ_c and the second increases. The first represents the decrease in burner temperature rise, for fixed θ_t, as τ_c is increased. The second represents the improving thermal efficiency of the cycle with increasing τ_c. Differentiating (2.7) with respect to τ_c, setting the result to zero, and solving for τ_c yields

$$\tau_c = \frac{\theta_t^{1/2}}{\theta_0} \qquad \text{(maximum } F/\dot{m}a_0\text{)}. \qquad (2.11)$$

The resultant value of $F/\dot{m}a_0$ is

$$(F/\dot{m}a_0)_{\max} = M_0(\{1 + [(\theta_t^{1/2} - 1)^2/(\theta_0 - 1)]\}^{1/2} - 1), \qquad (2.12)$$

and the corresponding specific impulse is

$$I = (a_0 h/g c_p T_0) \frac{M_0(\{1 + [(\theta_t^{1/2} - 1)^2/(\theta_0 - 1)]\}^{1/2} - 1)}{\theta_t - \theta_t^{1/2}}. \qquad (2.13)$$

There is no similar optimum value of τ_c insofar as the specific impulse is concerned. A little study of (2.8) shows that I increases monotonically as τ_c is increased to the value θ_t/θ_0 which just reduces the burner temperature rise to zero. Putting $\tau_c = (\theta_t/\theta_0)(1 - \varepsilon)$, expanding (2.8) in ε, and letting ε tend to zero shows that the limiting value is

$$I = (a_0 h/g c_p T_0)[M_0(\theta_t - 1)/2\theta_t(\theta_0 - 1)] \qquad (\tau_c \to \theta_t/\theta_0). \qquad (2.14)$$

Referring to our discussion in chapter 1, we note that this limit corresponds to a choice of cycle (see figure 1.5) in which point 3 approaches point 4, maximizing the compressor temperature ratio and hence the thermal efficiency, and at the same time driving the propulsive efficiency to unity. Unfortunately, the thrust is also zero. In any real cycle, losses would cause the

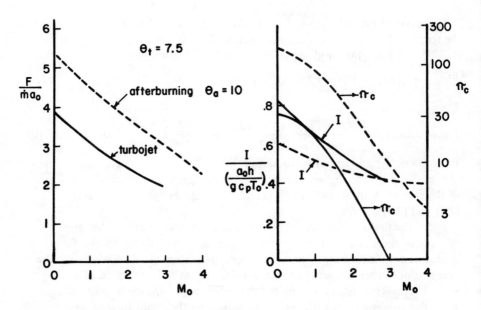

2.3 *Solid lines* indicate thrust per unit mass flow, specific impulse, and compressor pressure ratio for maximum thrust for simple turbojet with turbine inlet temperature 7.5 times ambient temperature. *Dashed lines* show afterburning turbojet with afterburner temperature 10 times ambient temperature.

thrust to be small long before τ_c approaches this limit. Nevertheless, the argument does serve to show that *the compressor pressure ratio that maximizes I is larger than the one that maximizes* $F/\dot{m}a_0$.

The results of (2.12), showing the maximum thrust per unit airflow that can be attained, are plotted as the solid lines in figure 2.3 for $\theta_t = 7.5$, which corresponds to a turbine inlet temperature of $1625°K$, and $T_0 = 216°K$ (stratosphere). The corresponding specific impulse, and the compressor pressure ratio required to achieve these values, are also given. Note that the (optimum) compressor pressure ratio drops rapidly, above $M_0 = .5$, and reaches 1.0 at $M_0 = 2.95$; at this Mach number the optimum turbojet has become a ramjet! Beyond $M_0 = 2.95$, the optimum π_c is less than unity; thus in this range the turbine should be before the combustor and the compressor after it. Note, however, that in this range of M_0 one should remove the limit on T_{t4} along with the turbomachinery, so that the calculation is really meaningless beyond the value of M_0 where $\tau_c = 1$.

2.4 The Afterburning Turbojet

Suppose that we now add to the turbojet a second combustion chamber, or afterburner, downstream of the turbine as indicated in figure 1.4. Temperatures higher than those in the combustor can be used here because the surface area is small enough to permit cooling and there are no moving blades exposed to the hot gas. Let T_{t7}/T_0 be denoted θ_a. Retracing the cycle analysis we find that $F/\dot{m}a_0$ is modified only by replacing τ_b by $\tau_b\tau_a$ where $\tau_a = T_{t7}/T_{t5}$. It is more useful, however, to have the expression explicit in terms of θ_a. Thus, writing $\tau_b = \theta_t/\theta_0\tau_c$ and $\tau_a = \theta_a/\theta_t\tau_t$ and substituting in (2.7), we find for the afterburning turbojet,

$$F/\dot{m}a_0 = M_0[([\theta_a/(\theta_0 - 1)]\{1 - (\theta_t/\theta_0\tau_c)/[\theta_t - \theta_0(\tau_c - 1)]\})^{1/2} - 1]. \quad (2.15)$$

From an energy balance across the entire engine, we see that

$$(\dot{m}_f + \dot{m}_a)h = \dot{m}c_p(T_{t7} - T_{t0});$$

thus $\dot{m}_f + \dot{m}_a = \dot{m}(c_pT_0/h)(\theta_a - \theta_0)$, and the specific impulse is simply

$$I = (a_0h/gc_pT_0)\frac{M_0\{[[\theta_a/(\theta_0 - 1)](1 - \{(\theta_t/\theta_0\tau_c)/[\theta_t - \theta_0(\tau_c - 1)]\})]^{1/2} - 1\}}{\theta_a - \theta_0}.$$

$$(2.16)$$

Differentiating (2.15) with respect to τ_c, we find that the τ_c that maximizes $F/\dot{m}a_0$ is

$$\tau_c = \tfrac{1}{2}(1 + \theta_t/\theta_0), \quad (2.17)$$

and the resultant maximum $F/\dot{m}a_0$ is

$$(F/\dot{m}a_0)_{max} = M_0[([\theta_a/(\theta_0 - 1)]\{1 - [4\theta_t/(\theta_0 + \theta_t)^2]\})^{1/2} - 1]. \quad (2.18)$$

We note first that the optimum τ_c given by (2.17) is larger than that for a nonafterburning turbojet, given by (2.11). The optimum values of $\pi_c = \tau_c^{\gamma/(\gamma-1)}$ are shown in figure 2.3 (dashed lines) in comparison to those for the simple turbojet. For $M_0 = 2$, the optimum π_c for the afterburning turbojet is about 30, while for the simple turbojet it is only about 4. While the simple cycle analysis exaggerates this difference, it is true that for supersonic flight the optimum compressor pressure ratio is higher with afterburning then without. This is fortunate, for a given engine with fairly high compression ratio can operate almost optimally with afterburning at M_0 near 2, where maximum thrust is required, and without afterburning subsonically, where the thrust demand is lower and good specific impulse is desired.

For the SST (supersonic transport), cruise at $M_0 > 2$ is the most important engine requirement. Our calculation indicates that an afterburning turbojet with $\pi_c \approx 12$ should have as good specific impulse as an optimum simple turbojet (with $\pi_c \approx 1.4$), and about 50 percent higher thrust for a given airflow. Considering the much better subsonic performance of the higher pressure ratio engine, it is clearly the best choice. The General Electric GE-4 afterburning engine which was under development for the SST had a pressure ratio of 12 at takeoff and somewhat less at cruise condition. The Concorde also uses an afterburning turbojet.

The results given in figure 2.3 portray the behavior of a family of engines, each having the optimum compressor pressure ratio for the particular Mach number. Suppose now that we fix the compressor pressure ratio, thus more nearly (but still not accurately) approximating the behavior of a fixed engine design as the Mach number varies. Thrust and specific impulse are given for a family of turbojets with $\pi_c = 6, 12, 24$ in this way in figure 2.4. Below $M_0 = 1.5$, the high pressure ratio engine is superior in both thrust and specific impulse. Above $M_0 = 1.5$, there is a tradeoff; low π_c is better for thrust and high π_c is better for specific impulse. The small circles on the curves of $F/\dot{m}a_0$ mark the points where, according to figure 2.3, the engine with fixed pressure ratio is optimum for thrust production. They would, of course, form an envelope, which is the curve of $F/\dot{m}a_0$ given in figure 2.3.

The thrust per unit airflow of the simple turbojet falls off badly at high M_0 because the compressor outlet temperature rises to meet the turbine inlet temperature, limiting the temperature rise of the combustor. An afterburner improves this situation greatly, as may be seen from figure 2.5 where the variation of $F/\dot{m}a_0$ and I with M_0 is compared for an engine with $\pi_c = 12$, with and without afterburner. The afterburning engine has about 30 percent more thrust at $M_0 = 0$ and maintains its thrust well up to $M_0 = 4$, where the simple turbojet has stopped functioning. The penalty in I shown for afterburning is realistic for $M_0 < 2$, but for M_0 near the value where $\theta_0 \tau_c \to \theta_t$ the specific impulse of a real simple turbojet would fall rapidly to zero, so

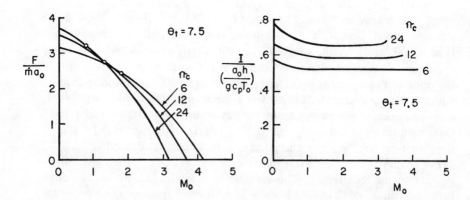

2.4 Thrust per unit mass flow and specific impulse for turbojet engines of various *fixed* compressor pressure ratios.

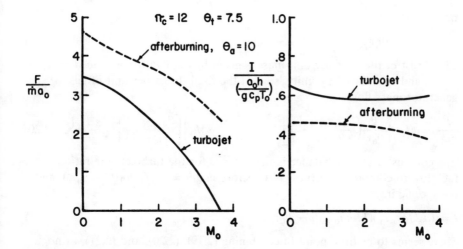

2.5   Comparison of thrust per unit mass flow and specific impulse of afterburning and simple turbojets, with fixed compressor pressure ratio, as functions of flight Mach number.

that the afterburning engine is in fact superior in specific impulse also for high M_0.

2.5   The Turbofan

In the turbofan engine shown schematically in figure 1.6, a part of the airflow through the fan is discharged directly through a nozzle to produce thrust; the remainder passes through the compressor, combustor, and turbine of the gas generator, then is exhausted through the primary nozzle. For convenience in the cycle analysis, we denote the overall compression ratio through fan and compressor by π_c, recognizing that this would in fact be the product of the fan and compressor pressure ratios. The pressure ratio of the fan alone will be π_f. A key parameter of the turbofan is the bypass ratio α, defined as the ratio of airflow through the bypass duct to that through the gas generator (compressor). As α is increased, more power is taken from the primary jet and put into the bypass jet.

Up to the point where the turbine temperature ratio was related to the compressor temperature ratio, the calculation of the turbojet's thrust applies equally well to the thrust of the primary jet of the turbofan, so that (2.6) applies directly. But now, to determine τ_t, we must take account of the power flowing to the fan. Thus,

$$\dot{m}c_p(T_{t4} - T_{t5}) = \dot{m}c_p(T_{t3} - T_{t2}) + \alpha\dot{m}(T_{t7} - T_{t2})c_p$$

or

$$\tau_t = 1 - [\theta_0/\theta_t][\tau_c - 1 + \alpha(\tau_f - 1)]. \tag{2.19}$$

The thrust of the duct per unit mass is given by (2.6) with τ_c replaced by τ_f, and τ_t and τ_b replaced by unity, so that the total thrust per unit of gas generator mass flow is

$$\frac{F}{\dot{m}a_0} = M_0\left\{\left[\frac{\theta_t}{\theta_0\tau_c}\left(\frac{\theta_0\tau_c\tau_t - 1}{\theta_0 - 1}\right)\right]^{1/2} - 1\right\} + \alpha M_0\left\{\left[\frac{\theta_0\tau_f - 1}{\theta_0 - 1}\right]^{1/2} - 1\right\}. \tag{2.20}$$

The combustor energy balance is the same as for the turbojet; the ratio of fuel flow to compressor airflow is as before $\dot{m}_f/\dot{m} = (c_pT_0/h)(\theta_t - \theta_0\tau_c)$, and the specific impulse is

$$I = (a_0h/gc_pT_0)(F/\dot{m}a_0)/(\theta_t - \theta_0\tau_c). \tag{2.21}$$

There seems to be little point in combining (2.19), (2.20), and (2.21), as no real simplification results. We are now faced with the fact that in addition to the parameters M_0, τ_c, and θ_t that characterized the turbojet, we have α and τ_f to consider.

It is a complex parametric problem to determine the best choice of θ_t, τ_c, α, and τ_f for a given application, but a little reflection will yield some simplifications and generalizations. We recall first that the basic reason for adding the fan to the turbojet is to improve the propulsive efficiency, that is, to make the jet velocity more nearly equal to the flight velocity. Now for a given total airflow, the propulsive efficiency will be highest if the duct and primary jets have the same velocity. (This follows from the variation of jet energy as the square of the velocity, and thrust directly as the velocity.) The number of parameters can therefore be reduced by considering only engines for which the two jet velocities are equal. In (2.20) this condition is equivalent to putting the two terms in square roots equal, subject to τ_t being given by (2.19). The result is the following expression for the fan temperature ratio:

$$\tau_f = \frac{1 + \theta_t + \theta_0(1 + \alpha - \tau_c) - (\theta_t/\theta_0\tau_c)}{\theta_0(1 + \alpha)} \tag{2.22}$$

Corresponding to this choice of fan temperature ratio, the thrust per unit of *total airflow* is

$$\frac{F}{\dot{m}a_0(1 + \alpha)} = M_0\left(\left\{\frac{\theta_t - (\theta_t/\theta_0\tau_c) - \theta_0(\tau_c - 1) + \alpha(\theta_0 - 1)}{(\theta_0 - 1)(1 + \alpha)}\right\}^{1/2} - 1\right). \tag{2.23}$$

One of the most important applications of turbofans is to high-subsonic-cruise transport aircraft. Here the fuel consumption in cruise is a major consideration, but engine weight is also important. To see how these criteria interact, we may plot the thrust per unit airflow and specific impulse from

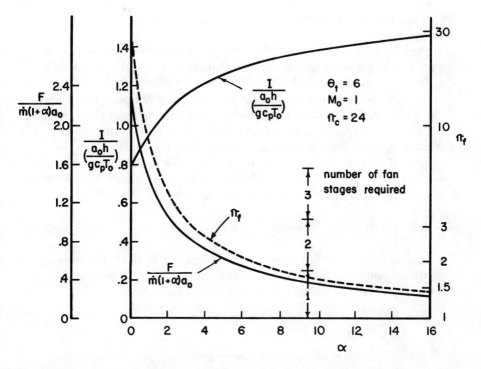

2.6 Thrust per unit of total mass flow and specific impulse for turbofan engines as functions of bypass ratio α at flight Mach number $M_0 = 1$, showing also the required fan pressure ratio.

(2.23) as functions of bypass ratio, for fixed θ_t and τ_c and for $M_0 = 1$. Such plots are given in figure 2.6 for $\theta_t = 6$ and $\pi_c = 24$; both values are representative of the engines for the wide-bodied transports of the early 1970s. They show that the thrust per unit of airflow drops off rapidly as α increases from zero, but there is a very significant improvement in I at the same time. A good part of this improvement is achieved by increasing α to 5, the value selected for modern large commercial transport engines such as the Pratt & Whitney JT9-D and the General Electric CF-6. Our plot shows that these engines should have fan pressure ratios near 2.5. As indicated, with current technology this would necessitate two fan stages. The actual engines have one stage to reduce fan noise and weight. A military engine, the GE TF-39 for the C5A heavy logistics transport, has a bypass ratio of 8, and "one and one-half" fan stages.

In general the optimum bypass ratio for any particular application is determined by a tradeoff between fuel weight, reduced by increasing α, and engine weight, which increases with α for a given thrust.

2.6 The Afterburning Turbofan

From section 2.5 it is clear that the turbofan is an excellent subsonic cruise engine. It also has much to offer aircraft that must cruise subsonically and also fly supersonically, provided an afterburner is added.

Ordinarily, the gas generator and fan airflows are mixed prior to entrance to the afterburner, as indicated in figure 2.7, and burning takes place in the entire airflow. This imposes a condition on the fan pressure ratio, namely that $p_{t7} = p_{t5}$, or $\pi_f = \pi_c\pi_t$. For ideal components this implies that $\tau_f = \tau_c\tau_t$. Now, if burning raises the fan and gas generator exhaust streams to the same temperature at station 6, denoted by θ_a, the two streams will have the same exhaust velocity, and from (2.20) we may write the thrust simply as

$$\frac{F}{\dot{m}a_0} = (1 + \alpha)M_0\left(\left\{\left(\frac{\theta_a}{\theta_0\tau_f}\right)\left(\frac{\theta_0\tau_f - 1}{\theta_0 - 1}\right)\right\}^{1/2} - 1\right). \tag{2.24}$$

The turbine, compressor-fan power balance expressed by (2.19) must apply, but with the condition that $\tau_f = \tau_c\tau_t$. Eliminating τ_t from (2.19) with this expression gives the following relation for τ_f as a function of α:

$$\tau_f = \frac{\theta_t + \theta_0(1 + \alpha - \tau_c)}{\theta_t/(\tau_c + \alpha\theta_0)}. \tag{2.25}$$

Since the gas generator and duct streams come to the same temperature at station 6, the combustion energy balance is very simple,

$$\dot{m}_fh = \dot{m}(1 + \alpha)c_pT_0(\theta_a - \theta_0),$$

and the specific impulse is

$$I = (a_0h/gc_pT_0)M_0\{[(\theta_a/\theta_0\tau_f)(\theta_0\tau_f - 1)/(\theta_0 - 1)]^{1/2} - 1\}/(\theta_a - \theta_0). \tag{2.26}$$

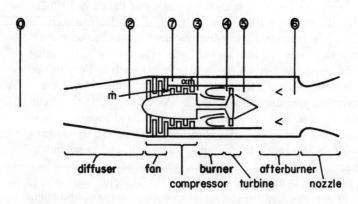

2.7   Schematic diagram of an afterburning turbofan engine.

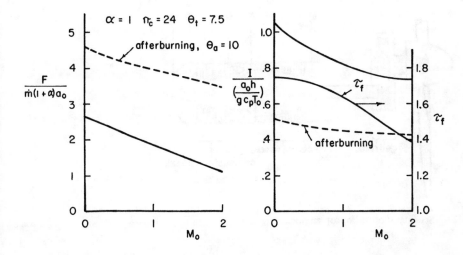

2.8 Comparison of thrust per unit of total mass flow and specific impulse for *afterburning* and *nonafterburning turbofans*, with fan temperature ratio required to match fan and turbine outlet pressures.

The performance of this engine, for $\alpha = 1$, $\pi_c = 24$, is shown in figure 2.8 along with its nonafterburning performance from (2.20) and (2.21). Several points should be noted. First, as indicated on the right figure, τ_f varies considerably with M_0 for the matching conditions imposed, namely $\pi_f = \pi_c \pi_t$ and $\alpha = 1$. This would imply for a fixed engine that π_f must decrease with increasing M_0, relative to π_c, a requirement that can be met only with some effort. In practice α would also vary somewhat with M_0, reducing the required variation of π_f. Second, the engine has a large ratio of afterburning to nonafterburning thrust. This can be advantageous for meeting dual requirements of subsonic cruise and supersonic dash. Third, the subsonic, nonafterburning specific impulse is considerably better than for a turbojet (compare figure 2.5), which is a further advantage for subsonic cruise. But, finally, these advantages are somewhat offset by a rather low afterburning specific impulse.

The engines for a number of modern high-performance military aircraft are of this type, including the F-111, and F-14 and the F-15. The Russian SST, the TU-144, also uses an afterburning turbofan.

2.7 The Turboprop

In a sense the turboprop shown schematically in figure 2.9 may be regarded as a turbofan engine with very high bypass ratio and correspondingly high

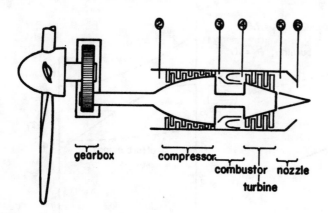

2.9   Schematic of turboprop engine.

propulsive efficiency. Practically speaking, however, there are very large qualitative differences in the two engines. The propeller is exposed to the oncoming flow without a diffuser, and this means that the tips of the propeller blades are exposed to a flow velocity that is the vector sum of the aircraft's flight velocity plus the rotational tip velocity. The result is that the tips reach sonic velocity at modest flight speeds. While in principle it is possible to build propellers that operate supersonically (in this sense), experience has shown that they are noisy and inefficient. The result is that in practice turboprop aircraft have been limited to flight Mach numbers not much above 0.5. Higher speeds may be possible with slow-turning, many-bladed propellers.

Another factor that leads to unique characteristics for the turboprop (among turbine engines) is the flexibility that results from variation of the pitch of the propeller. By varying the pitch, the efficiency of the propeller, defined as its propulsive power divided by the power supplied to it, can be made reasonably constant over a considerable speed range. (Of course this is not true at $M_0 = 0$, where this efficiency is necessarily zero.) This characteristic has led to the treatment of turboprops as devices for producing shaft power, rather than thrust, and this practice will be followed here with due regard for any jet thrust that may be produced in addition to the propeller power.

The total propulsive power of the turboprop is the sum of the jet propulsive power Fu_0 plus the shaft power P delivered to the propeller times the propeller efficiency. For the ideal engine, the propeller efficiency is taken to be unity. It is convenient to define a total work coefficient W as

$$W = [P/(\dot{m}c_p T_0)] + [(u_0 F)/(\dot{m}c_p T_0)], \tag{2.27}$$

that is, W is defined as the ratio of propulsive power to the thermal energy flow into the engine in the airflow.

The jet thrust is given by (2.6), with the appropriate value of τ_t to account for the propeller work; thus

$$W = (P/\dot{m}c_p T_0) + [(\gamma - 1)M_0^2]\{[(\theta_t/\theta_0\tau_c)(\theta_0\tau_c\tau_t - 1)/(\theta_0 - 1)]^{1/2} - 1\}.$$

Balancing the compressor and propeller powers with the turbine power gives

$$P + \dot{m}c_p(T_{t3} - T_{t2}) = \dot{m}c_p(T_{t4} - T_{t5})$$

or

$$\theta_t(1 - \tau_t) = \theta_0(\tau_c - 1) + P/\dot{m}c_p T_0.$$

Rather than substitute this result directly for τ_t, it is helpful to define a parameter,

$$\chi = (p_{t6}/p_0)^{(\gamma-1)/\gamma} = \delta_6^{(\gamma-1)/\gamma},$$

which characterizes the extent to which the available energy has been taken from the exhaust flow. Clearly $\chi > 1$ if the engine is to operate, since p_{t6} must be greater than p_0 for the exhaust gas to flow from the nozzle. Forming ratios of pressures through the engine gives

$$\chi = (\delta_0\pi_c\pi_t)^{(\gamma-1)/\gamma} = \theta_0\tau_c\tau_t;$$

so $\tau_t = \chi/\theta_0\tau_c$ and

$$P/\dot{m}c_p T_0 = \theta_t(1 - \chi/\theta_0\tau_c) - \theta_0(\tau_c - 1).$$

Substituting in the expression for W gives

$$W = \theta_t[1 - (\chi/\theta_0\tau_c)] - \theta_0(\tau_c - 1) + (\gamma - 1)M_0^2\left(\left\{\left(\frac{\theta_t}{\theta_0\tau_c}\right)\left(\frac{\chi - 1}{\theta_0 - 1}\right)\right\}^{1/2} - 1\right).$$

$$(2.28)$$

The rate of fuel consumption per unit of propulsive work is conventionally used to characterize the efficiency of the engine. Denoting this "specific fuel consumption" by $s = \dot{m}_f/\dot{m}c_p T_0 W$ and writing a heat balance for the combustor, $\dot{m}_f h = \dot{m}c_p T_0(\theta_t - \theta_0\tau_c)$, gives

$$s = (\theta_t - \theta_0\tau_c)/hW.$$

$$(2.29)$$

If the energy content of the fuel h is expressed in rational units, say Joules/kg, then s has units of kg/Joule. It is more conventional to use thermal units for h, for example BTU per lb, and to give s in "lb of fuel per horsepower hour." A little pondering should in this case lead to

$$s = 2545(\theta_t - \theta_0\tau_c)/hW, \quad \text{lb fuel/hp hr};$$

the dimensions of h are BTU/lb.

The choice of χ determines the distribution of power between jet and propeller. What value of χ maximizes W, for given values of M_0, τ_c, and θ_t? Differentiating W with respect to χ and setting the result to zero will show that χ^* is given by

$$\{(\theta_t/\theta_0\tau_c)[(\chi^* - 1)/(\theta_0 - 1)]\}^{1/2} = 1. \tag{2.30}$$

This is precisely the condition which makes $u_6/u_0 = 1$, the jet thrust zero, *and the overall propulsive efficiency of the engine unity.* Because the propeller has been assumed to be perfectly efficient, it is best to shift all available energy from the jet to the propeller. If an efficiency $\eta_{\text{propeller}}$ is assigned to the propeller, the optimum χ^* is given by

$$\{(\theta_t/\theta_0\tau_c)[(\chi^* - 1)/(\theta_0 - 1)]\}^{1/2} = 1/\eta_{\text{propeller}}.$$

If we retain the simple result (2.30), the maximum value of W and the corresponding s are

$$W^* = (\theta_t - \theta_0\tau_c)\left(\frac{\theta_0\tau_c - 1}{\theta_0\tau_c}\right), \tag{2.31}$$

$$hs^* = \frac{\theta_0\tau_c}{\theta_0\tau_c - 1}. \tag{2.32}$$

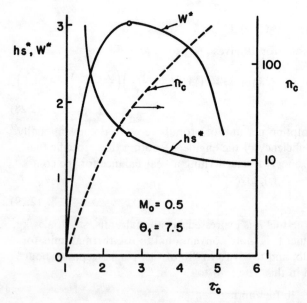

2.10 Propulsive work per unit of inlet enthalpy flux and specific fuel consumption for turboprop engine, as functions of compressor temperature ratio. (Multiply hs^* by $2545/18,500 = 0.14$ to find s in lb/hp hr.)

These results are plotted in figure 2.10 as functions of τ_c for $M_0 = 0.5$ and $\theta_t = 7.5$. A sharp peaking of the power is noted, with a continuous decrease of the specific fuel consumption. The inclusion of losses would of course cause s^* to turn up again at large τ_c.

Differentiating (2.31) with respect to τ_c shows that the value of τ_c that maximizes W^* is

$$\tau_c = \theta_t^{1/2}/\theta_0 \quad \text{(maximum } W^*\text{)},$$

the same as for the simple turbojet. The peak W^* is

$$(W^*)_{\text{max}} = (\theta_t^{1/2} - 1)^2 \tag{2.33}$$

and

$$h(s^*)_{\text{max}} = \theta_t^{1/2}/(\theta_t^{1/2} - 1). \tag{2.34}$$

These points are indicated on figure 2.10 by the circles.

Evidently, for $M_0 < 0.5$ these results do not depend strongly on M_0, as only $\theta_0 = 1 + \frac{1}{2}(\gamma - 1)M_0^2$ enters. They depend mostly on θ_t and τ_c. Further, since the jet thrust has been set to zero, they apply to turboshaft engines, such as might power helicopters and trains.

2.8 The Regenerative Gas Turbine

An important modification of the turboprop (or turboshaft) engine is the addition of a regenerator. This is a heat exchanger that withdraws heat from the turbine exhaust gas and adds it to the compressor outlet air ahead of the combustor, as indicated in figure 1.8. At low compressor pressure ratios, regeneration yields large reductions in specific fuel consumption. It is therefore very useful in automotive gas turbines, for example, where cost militates against an efficient high pressure ratio compressor. However, its weight has so far ruled it out for aircraft engines, where a high compression ratio is preferred.

To estimate the performance characteristics of the ideal regenerative engine, we note first that an ideal regenerator would raise the compressor outlet airflow from T_3 to $T_{3'}$ of figure 1.9 by transferring heat from the turbine exhaust, thus lowering its temperature from T_6 to $T_{6'}$. If the turbine exhaust gases had equal specific heat and mass flow to the compressor outlet air, $T_{6'}$ would then equal T_3. Actually $T_{3'}$ must always be below T_6; the difference is determined by the regenerator "effectiveness." For the ideal regenerator, we have

$$T_{3'} = T_6 = T_0\theta_t\tau_t;$$

the only modification of the analysis of the turboprop is in the combustor heat balance, which reads $\dot{m}_f h = \dot{m} c_p T_0 (\theta_t - \theta_t \tau_t)$, so that the specific fuel comsumption is

$$s = \frac{\theta_t (1 - \tau_t)}{hW} = \frac{\theta_t}{hW}\left(1 - \frac{\chi}{\theta_0 \tau_c}\right), \tag{2.35}$$

where W is again given by (2.28).

Since most applications of regenerative gas turbines are stationary or in low-speed vehicles, we can specialize (2.29) and (2.35) to $M_0 = 0$, whence

$$W = \theta_t (1 - \chi/\tau_c) - (\tau_c - 1),$$

$$s = \frac{\theta_t (1 - \chi/\tau_c)}{h[\theta_t (1 - \chi/\tau_c) - (\tau_c - 1)]}.$$

From (2.30), we see that $\chi^* \to 1$ as $\theta_0 \to 1$ for any value of $\theta_t/\theta_0 \tau_c$. This is reasonable, since it corresponds to $p_{t6}/p_0 = 1$, that is, to the exhaust total pressure equaling ambient pressure. With this condition we have finally

$$W^* = (\theta_t - \tau_c)(\tau_c - 1)/\tau_c, \tag{2.36}$$

which is identical to (2.31) for $\theta_0 = 1$, and

$$s^* = \theta_t / h(\theta_t - \tau_c). \tag{2.37}$$

Now the thermal efficiency η^*_{thermal} is $\eta^*_{\text{thermal}} \equiv P/\dot{m}_f h$; in the present case this becomes simply

$$\eta^*_{\text{thermal}} = \frac{1}{s^* h} = \frac{\theta_t - \tau_c}{\theta_t} = 1 - \frac{T_{t3}}{T_{t4}}. \tag{2.38}$$

Thus as $\tau_c \to 1$ the efficiency of the regenerative engine approaches

$$(\theta_t - 1)/\theta_t = 1 - T_0/T_{t4},$$

which is the Carnot efficiency for the temperature limits T_{t4} and T_0. Of course, from (2.36) the work per unit mass of air goes to zero in this limit; Figure 1.9 shows this happens where $T_{3'} \to T_4$ and $T_{6'} \to T_0$. Thus all heat is added at T_4 and rejected at T_0.

If we choose τ_c to maximize work for any given level of θ_t, we find $\tau_c = \theta_t^{1/2}$, and for this value

$$(W^*)_{\max} = (\theta_t^{1/2} - 1)^2,$$

$$h(s^*)_{\max} = \frac{\theta_t^{1/2}}{(\theta_t^{1/2} - 1)}, \qquad (\text{maximum } W)$$

exactly as for the nonregenerated engine (see (2.33) and (2.34)). In fact, for this condition of maximum work, $T_3 = T_6$ and there is no regeneration.

Thus we see that for the regenerated engine, as for the simple gas turbine, there is a tradeoff between power and efficiency; but for the regenerated engine, the efficiency is increased by lowering τ_c from the value for maximum power, while for the simple engine it was improved by increasing it from that for maximum power. In either case the efficiency approaches that of a Carnot cycle. The variations of W^* and hs^* for the simple and regenerative cycles as functions of compressor temperature ratio are shown in figure 2.11 for a value of $\theta_t = 6$, which is representative of modern (1970) turbines. We note that for low π_c (in the range of 5, such as is attainable with single stage centrifugal compressors) there is almost a factor of two reduction of fuel consumption with regeneration. To achieve the same s^* with the simple cycle would require $\pi_c \approx 60$, a high value indeed. Current compressor pressure ratios do not exceed about 30.

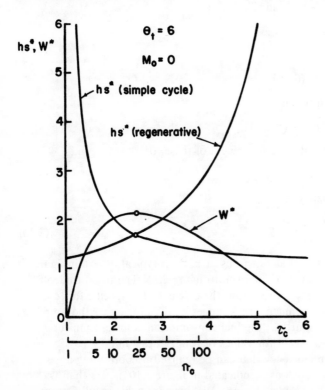

2.11 Comparison of specific work and specific fuel consumption of simple and regenerated shaft turbines.

2.9   Gas Turbines for Topping

When the gas turbine is used for "topping" a steam cycle, as shown schematically in figure 1.10, the idea is of course to recover some useful work from the heat rejected by the gas turbine. Since we propose to focus on the gas turbine here, it is actually more appropriate to think of the steam or other cycle as "bottoming" the gas turbine. From this viewpoint the ideal bottoming cycle is that which achieves the Carnot efficiency for each point along the heat rejection process, from 5 to zero. An elementary cycle of this sort is shown in figure 1.10. The power produced by this elementary cycle will be $dP = \dot{m}c_p[1 - T_0/T]\,dT$, where $\dot{m}$ is, as before, the mass flow in the gas turbine. The maximum total power from the bottoming cycle is then

$$P_{\text{bottom}} = \dot{m}c_p \int_{T_0}^{T_5} [1 - T_0/T]\,dT$$

$$= \dot{m}c_p T_0[T_5/T_0 - 1 - \ln{(T_5/T_0)}].$$

In the previous notation,

$$W_{\text{bottom}} = \theta_t \tau_t - 1 - \ln{(\theta_t \tau_t)}$$

$$= \theta_t/\tau_c - 1 - \ln{(\theta_t/\tau_c)}.$$

The work of the gas turbine is

$$W_{\text{gas turbine}} = \theta_t(1 - 1/\tau_c) - (\tau_c - 1).$$

The work of the combined cycle is the sum of these, or

$$W = \theta_t - \tau_c - \ln{(\theta_t/\tau_c)}, \tag{2.39}$$

while the thermal efficiency is

$$\eta = \frac{W}{\theta_t - \tau_c} = 1 - \frac{\ln{(\theta_t/\tau_c)}}{\theta_t - \tau_c}. \tag{2.40}$$

These relations are plotted as functions of τ_c for a typical $\theta_t = 6$ in figure 2.12, along with $hs = 1/\eta$ for comparison to figure 2.11. The usual tradeoff between power and efficiency is seen, but the efficiency (or specific fuel consumption) does not vary as much with changes of τ_c as for either the simple or the regenerated cycle. Further, the fuel consumption is lower than for either of the basic cycles except for values of τ_c very near unity.

Of course, the gain in efficiency that can be realized by bottoming a gas turbine with a real cycle, such as is indicated in figure 1.10, is less than we have found, because even an ideal vapor cycle does not achieve full Carnot efficiency and so recovers only a part of the available energy in the gas turbine exhaust. But when cycle comparisons are made with realistic estimates

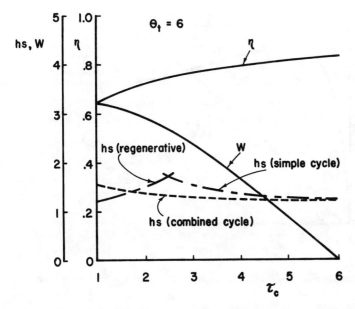

2.12 Efficiency η and specific work W of gas turbine with ideal (Carnot) bottoming cycle, with specific fuel consumption compared to that for simple and regenerated gas turbines.

of losses, this cycle retains its attractiveness. Indeed it is a major contender for application in modern, fossil-fueled power plants.

2.10 The Importance of Turbine Inlet Temperature

In the preceding discussion of various engines and their performance characteristics, the turbine inlet temperature T_{t4} or its dimensionless equivalent θ_t, has been assigned a "typical value" while the other cycle parameters such as τ_c and α, have been varied to illustrate the effect of such variation on the performance and to show how optimum values of these parameters may be determined. θ_t was held constant because performance of the gas turbine improves continually with θ_t; thus the maximum value, which turbine materials and cooling will allow, is usually selected. In fact, the steady improvement in gas turbine performance is to a considerable extent paced by improvements in oxidation resistant alloys of nickel and cobalt and, more recently, by the development of air-cooled turbine blades. The factors that limit this latter important development are discussed in some detail in chapter 6. Our purpose here is to display the trends of performance with turbine inlet temperature.

Beginning with the simple turbojet, suppose we consider the variations of

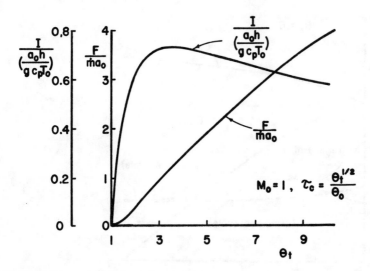

2.13   The effect of turbine inlet temperature ratio on thrust and specific impulse for the simple turbojet.

specific thrust $F/\dot{m}a_0$ and specific impulse with θ_t for a family of engines in which the compressor pressure ratio varies with θ_t to maximize $F/\dot{m}a_0$. This choice is arbitrary, but experience shows that real engines do not deviate greatly from it.

Figure 2.13 shows that for $M_0 = 1$, $F/\dot{m}a_0$ rises almost linearly with θ_t, while I first rises steeply, then gradually decreases in the range of θ_t from 4 to 10. The variation of I may be understood as a result of two competing trends. As θ_t increases, the thermal efficiency of the engine increases continuously while the propulsive efficiency decreases due to the increase in jet velocity (which leads to the increase in $F/\dot{m}a_0$). Thus, over the interesting range of θ_t, the specific impulse of a simple turbojet would deteriorate with increased turbine inlet temperature, although the dramatic increase in thrust per unit of airflow might still justify such an increase.

With the turbofan we can have increased thrust and also improved specific impulse, since the jet velocity can be lowered by increasing the bypass ratio for a given θ_t. To see this we return to (2.23) and take $\tau_c = \theta_t^{1/2}/\theta_0$, as for the turbojet. In (2.23) the square root represents the ratio of jet velocity to flight velocity, u_e/u_0. Suppose we hold u_e/u_0 constant as we vary θ_t. We find that the bypass ratio must then vary according to

$$1 + \alpha = \frac{(\theta_t^{1/2} - 1)^2}{(\theta_0 - 1)[(u_e/u_0)^2 - 1]},\tag{2.41}$$

which is obtained by equating the square root in (2.23) to u_e/u_0. The thrust

per unit of total airflow is simply

$$\frac{F}{\dot{m}a_0(1 + \alpha)} = M_0(u_e/u_0 - 1).$$ (2.42)

From (2.21), the specific impulse is

$$I = \left(\frac{a_0 h}{g c_p T_0}\right)\left[\frac{2}{(\gamma - 1)M_0}\right]\left[\frac{\theta_t^{1/2} - 1}{\theta_t^{1/2}}\right]\left[\frac{1}{u_e/u_0 + 1}\right].$$ (2.43)

(2.43) shows clearly the effects of flight Mach number, turbine inlet temperature, and jet velocity on the specific impulse.

The effect of θ_t alone is displayed in figure 2.14 by plotting the ratios of I
and $(1 + \alpha)$ to their values for a nominal θ_t of 7.5. I increases continually
with increasing θ_t, reflecting the improving thermal efficiency of the cycle,
for now η_p is constant since u_e/u_0 is held constant. The increased power of
the core engine is absorbed by a larger fan mass flow, as indicated by the
increasing of $(1 + \alpha)$ with θ_t.

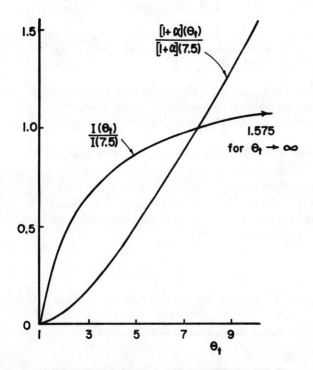

2.14   The effect of turbine inlet temperature ratio on specific impulse of turbofan engines
of constant propulsive efficiency. Also shown is the bypass ratio required to give the
constant propulsive efficiency (see equation (2.43)).

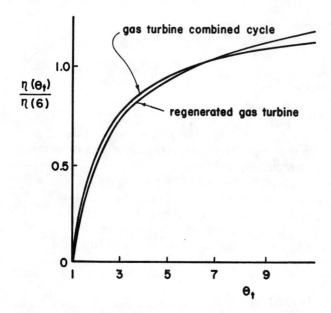

2.15   The effect of turbine inlet temperature ratio on efficiency of regenerated and ideally bottomed gas turbines (gas turbine combined cycle).

A reasonable upper limit for θ_t is about 10; for this value we might realize a 7 percent increase in I over the value for $\theta_t = 7.5$, which represents current (1976) advanced technology. For long-range aircraft this is a significant improvement.

There are similar trends for stationary (or automotive) gas turbines. Thus, from (2.38) and (2.40) we find the variations of η (relative to their values for $\theta_t = 6$) for regenerative gas turbines and gas turbine topping cycles, shown in figure 2.15. Not surprisingly, the trends are similar. Equally important, the power per unit of airflow (hence, roughly, for a given compressor size and cost) increases almost linearly with θ_t.

Problems

2.1 Using the results of figure 2.1, compute the thrust (in Newtons) of an ideal ramjet with a combustor inlet cross-sectional area of 1 m² and a combustor inlet Mach number $M_2 = 0.2$, at an altitude $h = 10$ km. Plot F versus M_0.

2.2 Very often air is "bled" from the compressor outlet of an aircraft engine for purposes such as deicing and boundary-layer control. Rederive the expressions for $F/\dot{m}a_0$ and $I/(a_0h/gc_pT_0)$ for a simple turbojet ((2.7) and

(2.8)) assuming a fraction ε of the compressor airflow is bled off. Compute values for $\varepsilon = 0.1$, $M_0 = 1$, $\theta_t = 7.5$, $\pi_c = 24$ and compare to figure 2.4.

2.3 Consider two jets of mass flow rates $\dot{m}_1$ and $\dot{m}_2$, with jet velocities u_1 and u_2. If the sum of the jet kinetic powers is fixed at a value P, show that the sum of the jet thrusts is maximum for $u_2 = u_1$.

2.4 One of the serious problems of high-speed flight is heating of the aircraft structure. Consider the following scheme for cooling it: Ram air is expanded through a turbine and cooled in the process. It is then used to cool the aircraft structure and heated in the process; then it is compressed by a compressor driven by the turbine and discharged through a nozzle. Carry out an ideal cycle analysis, assuming that the air is heated to T_{t0} before entering the compressor. You should find

$$F/\dot{m}u_0 = M_0(\{[\theta_0\tau_c/(\theta_0 - 1)] - [1/(\theta_0 - 1)(2 - \tau_c)]\}^{1/2} - 1).$$

Find the τ_c that maximizes $F/\dot{m}u_0$ and plot $(F/\dot{m}u_0)_{max}$ as a function of M_0. Can this device serve as the sole propulsion for a vehicle in steady flight?

2.5 Using the results of figure 2.4, compute the thrust (in Newtons) of an ideal turbojet with a compressor inlet area of 1 m^2, $\pi_c = 12$, and a Mach number at compressor inlet $M_2 = 0.5$. Plot F versus M_0 for $h = 0$ and for $h = 10$ km.

2.6 A turboprop engine with $\theta_t = 6$ and $\tau_c = 2$ is to fly at $M_0 = 0.8$. According to the ideal cycle analysis, what is the optimum core jet Mach number if the propeller efficiency is 1.0? if it is 0.8?

2.7 For an ideal turbojet engine with fixed θ_t and π_c, does the nozzle-exit static temperature increase or decrease with increasing M_0? What does this imply about the thermal efficiency of the engine?

2.8 In the afterburning turbofan discussed in section 2.6, the fan and core flows are mixed before afterburning. An alternative is to "afterburn" only in the fan airflow and exhaust the fan and core flows through separate nozzles. Carry out an ideal cycle analysis for such an engine, choosing the fan pressure ratio to make the fan and core exit velocities equal.

2.9 For the ideal ramjet, show that for fixed θ_b, $F/\dot{m}a_0$ is a maximum for $\theta_0 = \theta_b^{1/3}$.

3
Quantitative Cycle Analysis

The most important deviations from the ideal behavior described in chapter 2 occur as a result of

imperfect diffusion of the free-stream flow from flight to engine-inlet conditions;

nonisentropic compression and expansion in the compressor and turbine;

incomplete combustion and stagnation pressure loss in the burners;

incomplete expansion (or overexpansion) to ambient pressure in the nozzle; and

variation of the gas properties through the engine due to temperature changes.

Nozzle losses due to under- or overexpansion can be eliminated by design for proper expansion, but many engines use simple convergent nozzles for simplicity and weight savings. This particular loss, which is really a penalty due to fixed geometry and qualitatively different from the other nonidealities listed, will be included in the cycle analysis even though it would be more logical to consider it when the other ramifications of engine geometry are considered.

The aim at this point is to characterize each of these mechanisms for deviation from ideality so that their effects can be included in a more realistic cycle analysis. More detailed discussion of the sources of the losses and means for minimizing them will follow in subsequent chapters.

3.1 Variation in Gas Properties

As the air temperature rises in the compressor and as combustion changes the molecular composition of the gas as well as its temperature, its thermodynamic properties change. In the compressor the specific heat c_v rises with increasing temperature so that $\gamma = c_p/c_v$ decreases. A larger change occurs in the burner because of the large temperature rise and addition of polyatomic gases such as CO_2 and H_2O which have low values of γ.

It is possible to use tabulated values of the thermodynamic properties of air and of combustion gases and thus take accurate account of these effects in the engine cycle analysis; this method must be followed in computing performance of actual engines to the highest possible accuracy. A different procedure will be followed here to retain as much as possible the simplicity of the ideal cycle analysis. A mean c_{pc} and γ_c will be defined for the compressor and another pair of values, c_{pt} and γ_t, for the turbine. The first pair of values will be used for all processes occurring in the air ahead of the burner. The second set will be used for all processes in the combustion gases downstream of the burner. A mean specific heat $\bar{c}_p$ will be defined for the range of temperatures in the burner.

3.2 Diffuser Pressure Recovery

As the engine airflow is brought from the free-stream conditions ahead of the aircraft to the conditions required at entrance to the engine, it may be smoothly decelerated as in the subsonic inlet at the left in figure 3.1, or it may be decelerated through shock waves then further decelerated in a divergent passage as in the supersonic inlet shown at the right in the figure. In the subsonic inlets, viscous shear on the wall results in the growth of boundary layers that for this purpose may be thought of as regions of low stagnation pressure fluid. Mixing this fluid with the inviscid core flow results in some average reduction in stagnation pressure, below the value p_{t0} of the free

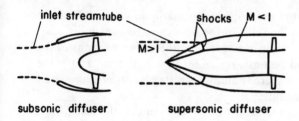

3.1 Schematic diagrams of diffusers, subsonic at left, supersonic at right.

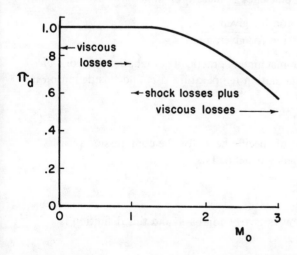

3.2 Typical diffuser pressure recovery as function of flight Mach number.

stream. The ratio of this average stagnation pressure at entrance to the engine, denoted p_{t2}, to the free-stream value will be termed the diffuser pressure recovery π_d. Thus

$$\pi_d = p_{t2}/p_{t0} = p_{t2}/p_0\delta_0.$$

When the flight velocity is supersonic, a further mechanism for stagnation pressure loss is the compression through a series of shocks. Such losses vary markedly with M_0 and for $M_0 > 2$ constitute the principal source of diffuser pressure drop.

A typical variation of π_d with M_0 is shown in figure 3.2 for $0 < M_0 < 3$. For $M_0 > 1$, it includes the loss through a single oblique shock and a normal shock, as sketched in figure 3.1.

3.3 Compressor and Turbine Efficiencies

Losses in compressors and turbines originate primarily in viscous shear regions on the blades and on the walls of the flow passages; these regions represent flows of lower stagnation pressure than the inviscid flow as in the diffuser. The low-energy fluid becomes mixed into the base flow, and at the compressor (or turbine) outlet there is an average stagnation pressure and an average stagnation temperature.

For a given stagnation pressure ratio from inlet to outlet, the result of losses in a compressor is to require more energy input than for an ideal (isentropic) compressor. The efficiency is therefore defined as

$$\eta_c = \frac{\text{ideal work of compression for given } \pi_c}{\text{actual work of compression for given } \pi_c}.$$

Because the flow through the machine is practically adiabatic, the work of compression all appears as stagnation temperature rise, and this definition is equivalent to

$$\eta_c = \{\pi_c^{(\gamma_c-1)/\gamma_c} - 1\}/(\tau_c - 1) \tag{3.1}$$

where now γ_c is a mean ratio of specific heats for the compression process.

Similarly, the turbine efficiency is defined as

$$\eta_t = \frac{\text{actual work for given } \pi_t}{\text{ideal work for given } \pi_t}.$$

For uncooled turbines the flow is nearly adiabatic and this definition is equivalent to

$$\eta_t = (1 - \tau_t)/[1 - \pi_t^{(\gamma_t-1)/\gamma_t}]. \tag{3.2}$$

Some modifications to this relation are in order for cooled turbines in which cooler air is introduced to cool the blades, but discussion of this rather complex subject will be deferred to chapter 6.

These definitions of efficiency are *not* the ones used in discussing steam turbines, for example, where the ratio of turbine output to inlet enthalpy in the steam above condenser conditions is taken as turbine efficiency. This is an energy extraction efficiency. η_c and η_t as defined here are measures of the approach of the actual processes to isentropic processes.

Apropos of the necessity for consistency between assumptions, it must be noted that there are relationships between π_c and η_c and between π_t and η_t. As will be explained in chapter 5, what is nearly constant between compressors of different π_c is the efficiency for a small pressure change and correspondingly small temperature change. This is termed the "polytropic efficiency" and will be denoted η_{pol}. Writing (3.1) for small changes in p_t and T_t gives

$$\eta_{pol} = \frac{(1 + \Delta p_t/p_t)^{(\gamma-1)/\gamma} - 1}{(1 + \Delta T_t/T_t - 1)} \approx \frac{\gamma - 1}{\gamma} \frac{\Delta p_t/p_t}{\Delta T_t/T_t} = \frac{\gamma - 1}{\gamma} \frac{d \log p_t}{d \log T_t}.$$

Now integrating and using the result to eliminate τ_c from (3.1) gives

$$\eta_c = \{\pi_c^{(\gamma_c-1)/\gamma_c} - 1\}/\{\pi_c^{(\gamma_c-1)/\gamma_c \eta_{pol}} - 1\};$$

similarly from (3.2)

$$\eta_t = \{1 - \pi_t^{(\gamma_t-1)(\eta_{pol})/\gamma_t}\}/\{1 - \pi_t^{(\gamma_t-1)\gamma_t}\}.$$

Values of η_c and η_t are shown in figure 3.3 for $\eta_{pol} = .90$, which is typical of modern turbomachines. Note that $\eta_t > \eta_{pol}$ while $\eta_c < \eta_{pol}$.

3.4 Burner Efficiency and Pressure Loss

Two distinct losses occur in the burner, combustion inefficiency and pressure loss. The first implies an imperfect job of converting the available chemical energy of the fuel-air mixture to thermal energy. Some fuel may remain as soot particles, some as partially combusted carbon in the form of CO. This loss will be characterized by a burner efficiency η_b defined as the change in enthalpy flux from inlet to exhaust of the burner divided by the fuel flow times the energy content of the fuel.

$$\eta_b = \bar{c}_p[(\dot{m} + \dot{m}_f)T_{t4} - \dot{m}T_{t3}]/\dot{m}_f h. \tag{3.3}$$

One of the subtleties in the combustion process is the basis for defining h, conventionally termed the "heating value" of the fuel. It is defined as the heat released when the fuel in stoichiometric mixture with air is burned in a steady flow process (at constant pressure), then cooled to the original ambient temperature. The heating values so determined will differ by about 10 percent

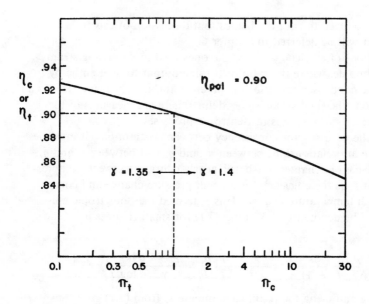

3.3 Compressor and turbine efficiencies as functions of pressure ratio for fixed polytropic efficiency.

depending upon whether the water vapor in the exhaust gas is condensed. If it is condensed, h includes its latent heat, and h is termed the "upper heating value." If the water leaves as vapor, h is termed the "lower heating value." This lower value is appropriate in gas turbines, since the water always leaves the burner as vapor. For a typical kerosene, h is about 4.30×10^7 Joule/kg. The upper heating value is nearer 4.65×10^7 Joule/kg.

The second loss is a drop in stagnation pressure, partially due to viscous effects and partially due to the stagnation pressure loss that occurs whenever heat is added to a flowing gas. The latter is not very important in turbine engine burners but plays a dominant role in supersonic combustion ramjet burners. The net effect of the two types of losses will be represented by a burner stagnation pressure ratio π_b.

3.5 Imperfect Expansion Loss

Perfect expansion of the exhaust nozzle implies that the flow is expanded isentropically to the final (ambient) pressure p_0. If the flow is underexpanded ($p_e > p_0$ in (1.12)), then further expansion occurs downstream of the nozzle, but since the flow has no surfaces to expand against, this expansion does not produce an increase in thrust over that represented by the momentum and pressure at the nozzle exit, and the thrust is reduced from that for $p_e = p_0$.

Physically, the value of p_e/p_0 is controlled by the nozzle geometry. Since no other component of the engine is described geometrically in the cycle analysis, it makes no sense to describe the nozzle in detail. Two limiting cases can be discussed, however. These are the ideal nozzle, which corresponds to $p_e = p_0$ and gives an upper limit on thrust, and the simple convergent nozzle, which gives a lower-limit for thrust. The exhaust nozzles of turbojets and even of turbofans are choked (they have pressure ratios larger than required to produce sonic velocity at the point of minimum area) under most flight conditions, and the condition at the exit plane of a convergent nozzle is then simply that $M_e = 1$, so this condition can be specified independent of the nozzle geometry. If the nozzle is not choked, the two limits coincide but $M_e < 1$.

3.6 Heat Exchanger Effectiveness and Pressure Loss

In the regenerative gas turbine of figure 1.8, the compressor discharge air is heated before entering the combustor by heat exchange with the turbine exhaust gas. This exchange would be carried out by either passing the air and exhaust gas alternately through a packed bed or honeycomb, as in figure 1.8, or by counterflow in a more conventional heat exchanger, where the hot low-pressure gases would be separated from the cool high-pressure air by metal tubes or plates.

Either case produces two kinds of deviation from ideal regeneration. First, the air and gas suffer stagnation pressure losses as they flow through the regenerator. The fractional pressure drop on the air side may be different from that on the gas side of the regenerator, but only the product of the two pressure ratios will appear in the cycle analysis. This product we denote π_r, so that in the notation of figures 1.8 and 1.9,

$$\pi_r = (p_{t6'}/p_{t6})(p_{t3'}/p_{t3}). \tag{3.4}$$

The regenerator would be thermally perfect if it raised the temperature of the compressor outlet air to the temperature of the turbine exhaust (if $T_{t3'} = T_{t6}$ in figure 1.9). Actually, there will always be some temperature drop across the heat exchanger plates or surfaces, so that $T_{t3'} < T_{t6}$. The effectiveness ε of the regenerator is defined as the ratio of the *actual* rise in compressor outlet air temperature to the ideal, or maximum possible. Thus,

$$\varepsilon = (T_{t3'} - T_{t3})/(T_{t6} - T_{t3}). \tag{3.5}$$

π_r and ε are related because the larger the flow Mach number through the regenerator, the larger the pressure drop for a given effectiveness. From section 4.1.4, we find

$$\pi_r = 1 - \alpha M_r^2[\varepsilon/(1 - \varepsilon)] \tag{3.6}$$

where α is a constant, best evaluated empirically but of order unity, and M_r is the Mach number of the flow through the regenerator. Clearly, the larger the flow passages for a given mass flow, the smaller M_r and the lower the pressure drop.

3.7 Turbojet with Losses

The simplest engine cycle that includes most of these effects is the turbojet cycle. Using the station numbers of figure 1.4, we find that the thrust is

$$F = \dot{m}_7 u_7 - \dot{m}u_0 + A_7(p_7 - p_0).$$

Let the overall fuel air ratio including primary combustor and afterburner be f; then $\dot{m}_7/\dot{m} = 1 + f$. The exit area can then be related to other parameters by conservation of mass:

$$A_7 = [\dot{m}(1 + f)/\rho_0 u_0](R_t/R_c)(p_0/p_7)(T_7/T_0)(u_0/u_7),$$

and the thrust may be written,

$$\frac{F}{\dot{m}u_0} = (1 + f)\frac{u_7}{u_0} - 1 + \frac{1 + f}{\gamma_c M_0^2}\frac{R_t}{R_c}\frac{T_7}{T_0}\frac{u_0}{u_7}\left(1 - \frac{p_0}{p_7}\right). \tag{3.7}$$

The task now is to determine the ratios T_7/T_0, p_7/p_0, u_7/u_0. Tracing the variations in stagnation temperature and pressure through the engine yields

$$T_{t7} = T_7[1 + \tfrac{1}{2}(\gamma_t - 1)M_7^2] = T_0\theta_0\tau_c\tau_b\tau_t$$
$$= T_0\theta_a \quad \text{(afterburning)} \tag{3.8}$$

and

$$p_{t7} = p_7\{1 + \tfrac{1}{2}(\gamma_t - 1)M_7^2\}^{(\gamma_t)/(\gamma_t - 1)} = p_0\delta_0\pi_d\pi_c\pi_b\pi_t\pi_a. \tag{3.9}$$

The first of these gives

$$\frac{T_7}{T_0} = \frac{\theta_0\tau_c\tau_b\tau_t}{1 + \tfrac{1}{2}(\gamma_t - 1)M_7^2} = \frac{\theta_a}{1 + \tfrac{1}{2}(\gamma_t - 1)M_7^2} \quad \text{(afterburning).} \tag{3.10}$$

The compressor-turbine power balance is

$$\dot{m}c_{pc}(T_{t3} - T_{t2}) = \dot{m}(1 + f_b)c_{pt}(T_{t4} - T_{t5})$$

where f_b is the fuel-air ratio of the primary burner. This can be written

$$\tau_t = 1 - [(c_{pc}/c_{pt})/(1 + f_b)][(\tau_c - 1)/(\tau_c\tau_b)]; \tag{3.11}$$

thus T_7/T_0 becomes

$$\frac{T_7}{T_0} = \frac{\theta_t - [(c_{pc}/c_{pt})/(1 + f_b)][\theta_0(\tau_c - 1)]}{1 + [(\gamma_t - 1)/2]M_7^2} = \frac{\theta_a}{1 + [(\gamma_t - 1)/2]M_7^2} \quad \text{(afterburning).} \tag{3.12}$$

Now $(u_7/u_0)^2 = (M_7/M_0)^2(\gamma_t R_t T_7/\gamma_c R_c T_0)$, and with the relations for T_7/T_0,

$$\frac{u_7}{u_0} = \left[\frac{\left[\left(\frac{c_{pt}}{c_{pc}}\right)\theta_t\left(1 - \frac{c_{pc}/c_{pt}}{1+f_b}\left(\frac{\theta_0(\tau_c - 1)}{\theta_t}\right)\right)\right]}{\theta_0 - 1} \frac{\frac{1}{2}(\gamma_t - 1)M_7^2}{1 + \frac{1}{2}(\gamma_t - 1)M_7^2}\right]^{1/2}$$

$$= \left[\frac{\left(\frac{c_{pt}}{c_{pc}}\right)\left(\frac{\theta_a}{\theta_0 - 1}\right)\frac{1}{2}(\gamma_t - 1)M_7^2}{1 + \frac{1}{2}(\gamma_t - 1)M_7^2}\right]^{1/2} \quad \text{(afterburning).}$$

(3.13)

From this point, it is convenient to recognize two limiting cases.

Convergent, choked nozzle. In this case, $M_7 = 1$, and u_7/u_0 is given directly by (3.13), while T_7/T_0 is determined by (3.12). The exit to entrance pressure ratio is controlled by the pressure balance through the engine. Thus, from (3.9),

$$p_7/p_0 = [(\gamma_t + 1)/2]^{-\gamma_t/(\gamma_t - 1)}\delta_0\pi_d\pi_c\pi_b\pi_t\pi_a.$$

From the definitions of η_c and η_t,

$$\pi_c = [1 + \eta_c(\tau_c - 1)]^{\gamma_c/(\gamma_c - 1)}$$

and

$$\pi_t = \{1 - [(1 - \tau_t)/\eta_t]\}^{\gamma_t/(\gamma_t - 1)}.$$

Thus

$$\frac{p_7}{p_0} = [(\gamma_t + 1)/2]^{-\gamma_t/(\gamma_t - 1)}\delta_0\pi_d\pi_c\pi_b\pi_a\left[1 - \frac{c_{pc}/c_{pt}}{1+f_b}\left(\frac{\theta_0(\tau_c - 1)}{\eta_t\theta_t}\right)\right]^{\gamma_t/(\gamma_t - 1)}. \quad (3.14)$$

There is little to be gained from substitution of these results into (3.7), so they will be left in this open form. Note that $F/\dot{m}a_0$ is a function only of M_0, τ_c, θ_t and θ_a.

Ideally expanded nozzle. For this limit, $p_7/p_0 = 1$, and the whole last term of (3.7) is zero, but now M_7 must be determined from the pressure balance (3.9). Thus

$$\frac{1}{2}(\gamma_t - 1)M_7^2 = (\delta_0\pi_d\pi_c\pi_b\pi_a)^{(\gamma_t - 1)/\gamma_t}\left[1 - \left(\frac{c_{pc}/c_{pt}}{1+f_b}\right)\left(\frac{\theta_0(\tau_c - 1)}{\theta_t\eta_t}\right)\right] - 1.$$

With (3.13) this determines $F/\dot{m}u_0$.

In either of the above two cases the specific impulse is given by

$$I = F/g\dot{m}_f = u_0(F/\dot{m}u_0)/gf \quad (3.15)$$

where $f = f_b + f_a$ is the total air-fuel ratio, f_b is the air-fuel ratio for the primary burner, and f_a that for the afterburner. Referring to the definition of burner efficiency, (3.3), gives

$$f_b = (\bar{c}_p T_0 / \eta_b h)[(1 + f_b)\theta_t - \theta_0 \tau_c] \qquad \text{(nonafterburning)}$$

$$f = f_b + f_a = (\bar{c}_p T_0 / \eta_b h)[(1 + f)\theta_a - \theta_0] \quad \text{(afterburning)}.$$

(3.16)

if the efficiencies of primary and afterburner are equal.

3.7.1 Effects of η_c, η_t, π_b, π_d

Reasonably accurate numerical estimates of thrust and specific impulse can be obtained from these results by straightforward computation. Before proceeding to this however, it is useful to ask what insight can be obtained from the explicit formulas by analytical manipulation. Consider first the effects of pressure losses, compressor efficiency, and turbine efficiency for ideal nozzle expansion. To eliminate some complexity, we put $c_{pc} = c_{pt}$, $\gamma_c = \gamma_t$, and $f_b \approx 0$. Equations (3.7) and (3.13) may then be combined to give

$$\frac{F}{\dot{m}u_0} + 1 = \left\{ \left(\frac{\theta_t - \theta_0(\tau_c - 1)}{\theta_0 - 1} \right) \right.$$

$$\left. \cdot \left(1 - \left[\theta_0 \pi_l^{(\gamma-1)/\gamma}[1 + \eta_c(\tau_c - 1)]\left(1 - \frac{\theta_0(\tau_c - 1)}{\eta_t \theta_t} \right) \right]^{-1} \right) \right\}^{1/2}$$

and

$$I = \frac{[(u_0 h \eta_b)/(\bar{c}_p T_0)](F/\dot{m}u_0)}{(1 + f_b)\theta_t - \theta_0 \tau_c}$$

where $\pi_l^{(\gamma-1)/\gamma} = (\pi_d \pi_b \pi_a)^{(\gamma-1)/\gamma}$.

Note first that if τ_c is held constant while η_c or η_t or π_l is varied then the variation of I is a constant times the variation of $F/\dot{m}u_0$. Letting $(F/\dot{m}u_0) + 1 = Z$, we find by differentiation that

$$\frac{\partial Z}{\partial \eta_c} = \frac{Q\tau_c}{1 + \eta_c(\tau_c - 1)}$$

$$\frac{\partial Z}{\partial \eta_t} = \frac{Q(\theta_0/\theta_t)[(\tau_c - 1)/\eta_t^2]}{1 - (\theta_0/\theta_t)[(\tau_c - 1)/\eta_t]}$$

(3.17)

$$\frac{\partial Z}{\partial \pi_l^{(\gamma-1)/\gamma}} = \frac{Q}{\pi_l^{(\gamma-1)/\gamma}}$$

where

$$Q = \frac{1}{2Z} \frac{\theta_t - \theta_0(\tau_c - 1)}{\theta_0(\theta_0 - 1)\pi_l^{(\gamma-1)/\gamma}[1 - \theta_0(\tau_c - 1)/\theta_t \eta_t][1 + \eta_c(\tau_c - 1)]}.$$

Some interesting points are immediately apparent. First, the relative sensitivity of Z to η_c and η_t is

$$\frac{(\partial Z/\partial \eta_c)}{(\partial Z/\partial \eta_t)} = \frac{[\tau_c/(\tau_c - 1)][\eta_t^2 \theta_t/\theta_0 - (\tau_c - 1)]}{1 + \eta_c(\tau_c - 1)}.$$

For typical values such as $M_0 = 1$, $\theta_t = 7.5$, $\tau_c = 2.5$, $\eta_c = \eta_t = .85$, this gives a value of about 3.3, showing that *the compressor efficiency is much more important than turbine efficiency in maximizing $F/\dot{m}u_0$ (and I).*

Forming a similar ratio of the sensitivities to $\pi_l^{(\gamma-1)/\gamma}$ and η_c gives

$$\frac{\{\pi_l^{(\gamma-1)/\gamma} \partial Z/\partial \pi_l^{(\gamma-1)/\gamma}\}}{(\partial Z/\partial \eta_c)} = \frac{\tau_c}{1 + \eta_c(\tau_c - 1)};$$

for the above typical values this is 1.1, so that a percent improvement in $\pi_l^{(\gamma-1)/\gamma} = (\pi_d \pi_b \pi_a)^{(\gamma-1)/\gamma}$ is as important as a percent improvement in η_c.

There are many possible uses of this type of analysis. Although the calculations tend to become complex algebraically, the added insight given by explicit expressions for the desired quantities often makes it more rewarding than direct numerical calculations.

Some typical numerical results for the turbojet with losses are presented in figure 3.4. In figure 3.4a the thrust per unit airflow for an engine with parameters as summarized in the figure is compared to that for the ideal engine, given by the circular points and by the square points for the afterburning engine. There is not much difference because the effects of reduced compressor and turbine efficiencies tend to cancel those of c_{pt} relative to c_{pc}. In an engine with a higher compressor pressure ratio, there would be a larger difference between the ideal and actual cycles.

The effect of the compressor and turbine efficiencies is more apparent in the specific impulse shown in figure 3.4b; the real cycle has about 10 percent lower values.

Nozzle performance is critical at the higher values of M_0; the lowest curve in figure 3.4a describes an engine with simple convergent nozzle. The loss in thrust (and in I) is about 60 percent at $M_0 = 3$. Note, however, that the loss due to imperfect expansion is small for $M_0 < 1$. Engines for subsonic aircraft characteristically have convergent nozzles for this reason.

3.8 Regenerative Gas Turbine with Losses

Shaft power P is the useful output of the regenerative gas turbine. From the compressor-turbine power balance

$$P = \dot{m}(1 + f_b)c_{pt}T_{t4}(1 - \tau_t) - \dot{m}c_{pc}T_0(\tau_c - 1)$$

where the notation is as in figure 1.8. In dimensionless form,

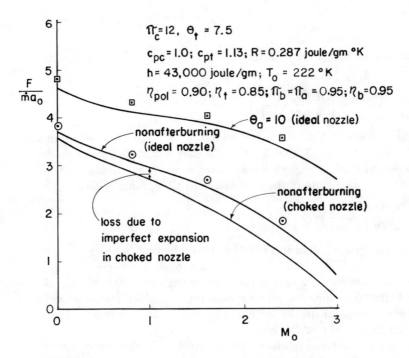

3.4a Thrust of turbojet with losses as a function of M_0, showing effects of nonideal nozzle expansion, and differences from ideal analysis, indicated by ▣ for afterburning and by ⊙ for nonafterburning.

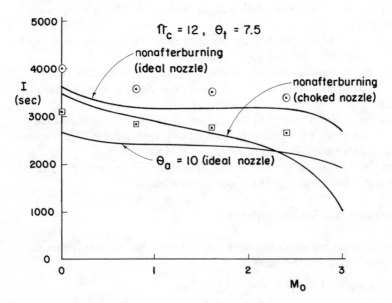

3.4b Specific impulse for conditions of figure 3.4a.

$$W = \frac{P}{\dot{m} c_{pc} T_0} = (1 + f_b) \frac{c_{pt}}{c_{pc}} \theta_t (1 - \tau_t) - (\tau_c - 1).$$

Since τ_c and θ_t are to be regarded as design parameters, we have only to determine τ_t (and f_b) in order to compute $P/\dot{m} c_p T_0$. Following the changes in stagnation pressure through the engine, as in (3.9), we find

$$p_{t6'} = p_0 \pi_c \pi_b \pi_t \pi_r,$$

where it will be noted that $\pi_r = (p_{t3'}/p_{t3})(p_{t6'}/p_{t6})$ includes the pressure losses in both sides of the regenerator.

The least exhaust loss will be achieved when $p_{t6'} = p_0$, when the exhaust gas can just flow out of the engine. With this condition, then,

$$\pi_t = (\pi_c \pi_b \pi_r)^{-1} \quad \text{and} \quad 1 - \tau_t = [-(\pi_c \pi_b \pi_r)^{-(\gamma_t - 1)/\gamma_t} + 1]\eta_t,$$

so that finally

$$W = (1 + f_b)(c_{pt}/c_{pc})\theta_t \eta_t [1 - (\pi_c \pi_b \pi_r)^{-(\gamma_t - 1)/\gamma_t}] - [\pi_c^{(\gamma_c - 1)/\gamma_c} - 1]\eta_c. \qquad (3.18)$$

The energy balance for the combustor gives

$$f_b = (\bar{c}_p T_0 / \eta_p h)[(1 + f_b)\theta_t - (T_{3'}/T_0)];$$

the definition (3.5) of regenerator effectiveness then gives

$$T_{3'}/T_0 = \varepsilon(\theta_t \tau_t - \tau_c) + \tau_c$$

so that

$$\frac{h f_b}{c_{pc} T_0} = \frac{\bar{c}_p}{\eta_p c_{pc}} \left[(1 + f_b)\theta_t - \varepsilon\theta_t \{1 - \eta_t[1 - (\pi_c \pi_b \pi_r)^{-(\gamma_t - 1)/\gamma_t}]\} \right.$$

$$\left. - (1 - \varepsilon)\left(1 + \frac{\pi_c^{(\gamma_c - 1)/\gamma_c} - 1}{\eta_c}\right) \right]. \qquad (3.19)$$

The thermal efficiency is now

$$\eta_{\text{thermal}} = \frac{P}{h f_b \dot{m}} = \frac{P/\dot{m} c_p T_0}{h f_b / c_{pc} T_0}; \qquad (3.20)$$

the numerator is given by (3.18) and the denominator by (3.19).

As for the turbojet not much is gained by further manipulation of (3.18) and (3.19). Computation is necessary to show the effects of the losses. The results of some typical calculations are given in figures 3.5 and 3.6. Figure 3.5 shows the effects of variations of regenerator pressure drop and effectiveness on efficiency and power. As expected, the optimum ε increases as αM_r^2 decreases, since π_r increases. For a typical value of $\alpha M_r^2 = 0.1$, the peak efficiency is about 0.42, while the power is reasonably close to the maximum of $P/\dot{m} c_{pc} T_0 = 0.3$.

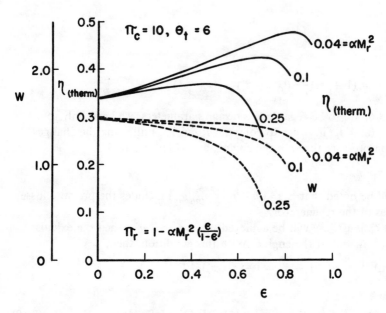

3.5   Efficiency $\eta_{(therm)}$ and specific work W of regenerated gas turbine with losses, as functions of regenerator effectiveness ε for various regenerator Mach numbers M_r.

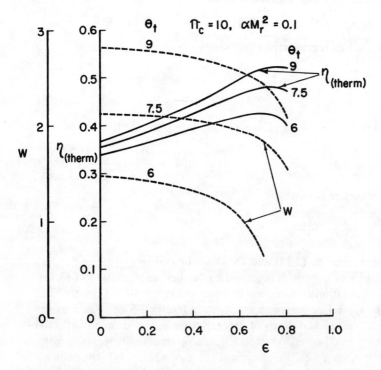

3.6   Effect of turbine inlet temperature ratio θ_t on efficiency and specific power of regenerated gas turbine.

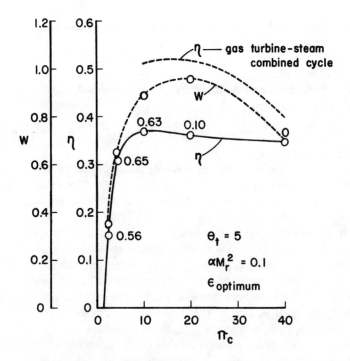

3.7   Efficiency and specific power of regenerated gas turbine, and gas turbine-steam combined cycle, as functions of compressor ratio. Regenerator effectiveness ε selected for maximum efficiency.

The effect of increasing θ_t with fixed $\pi_c = 10$ is shown figure 3.6. We see that the power increases dramatically, but the efficiency less so.

From a series of calculations such as those shown in figure 3.5, for different π_c the points of peak η can be selected, with the results summarized in figure 3.7. Here θ_t has been chosen as 5, a value representative of advanced, long-life, ground-based gas turbines, to facilitate comparison with the gas turbine–steam cycle in the next section. Note that for $\theta_t = 5$, a π_c of 10 maximizes η, and that the optimum ε at that point is 0.63. The power is somewhat below its maximum reached at $\pi_c \approx 20$, as would be expected.

3.9   Combined Gas Turbine–Steam Cycles with Losses

An assessment of this cycle, diagrammed in figure 1.10, necessitates a parametric study of both the steam "bottoming" cycle (a,b,c,d) and the gas turbine "topping" cycle $(0,3,4,5)$. The requirement that the steam cycle draw its heat from the gas turbine exhaust implies that the gas temperature along the line $5 \rightarrow 0$ is larger than the steam (or water) temperature along $a \rightarrow c$ at each set of points between which the fluids are exchanging energy. Some reflection will show that this implies

(1) $T_6 - T_b = \Delta T_b > 0$,

(2) $T_5 - T_c = \Delta T_s > 0$,

where ΔT_b and ΔT_s are the gas-water temperature drops in the boiler and superheater respectively. A further condition is

$$\dot{m}_{H_2O}(h_c - h_b) = \dot{m}_{gas}c_p(T_5 - T_6) \tag{3.21}$$

where h is the enthalpy of the steam.

Within these constraints is a wide choice of possible steam cycles, as well as gas cycles. Thus, given a maximum feasible value of T_4 and a heat rejection temperature $T_a = T_d$, we can choose a high steam pressure and temperature T_b with a correspondingly high value of T_6 or a relatively low T_b and T_6. The first choice would result in an efficient steam cycle and relatively inefficient gas cycle, while the second choice would reverse the situation.

To analyze this cycle, we begin by prescribing the two most important parameters of the steam cycle, namely the boiling temperature T_b (and hence the steam pressure p_b) and the superheat temperature T_c. These values determine the enthalpies h_c, h_b, h_a. If we specify the steam turbine efficiency, the steam turbine work is given by $h_c - h_{d'} = \eta_{t\,(steam)}[h_c - h_d]$ where h_d is the ideal value of enthalpy at the steam turbine exit, that is, the value for which the entropy at d equals that at c, and $\eta_{t\,(steam)}$ is the efficiency of the steam turbine. An example will clarify this calculation. We choose

$$p_a = p_d = p_{d'} = 2 \text{ mmHg}$$

$$\bar{p}_b = 2400 \text{ psia}$$

$$T_c = 1000\,^\circ F.$$

Then from reference 3.1 we find

$T_b = 662\,^\circ F$ $h_b = 718 \text{ BTU/lb}$

$h_c = 1462 \text{ BTU/lb}$ $s_c = 1.534 \text{ BTU/lb }\,^\circ F$

$T_a = 101\,^\circ F,$ $h_{af} = 69.1, h_{afg} = 1037 \text{ BTU/lb},$

$s_{af} = 0.132,$ $s_{afg} = 1.848 \text{ BTU/lb }\,^\circ F$

where the subscripts f and fg refer to the properties for the liquid and for vaporization, respectively. Equating the entropies at c and d for ideal expansion gives $s_c = 1.534 = s_{af} + x_d s_{afg} = 0.132 + 1.848 x_d$ or $x_d = 0.759$. Then $h_d = 69.1 + 0.759(1037) = 856$, and if $\eta_{t\,(steam)} = 0.85$, then $h_c - h_{d'} = 0.85(1462 - 856) = 515 \text{ BTU/lb}$. The ratio of steam mass flow to gas turbine mass flow follows from (3.21):

$$\dot{m}_{H_2O}\dot{m}_{gas} = \frac{c_{pt}(T_5 - T_6)}{h_c - h_b} = \frac{c_{pt}(T_c + \Delta T_s - T_b - \Delta T_b)}{h_c - h_b}.$$

For $\Delta T_s = 100°F$, $\Delta T_b = 50°F$,

$$\frac{\dot{m}_{H_2O}}{\dot{m}_{gas}} = \frac{0.27(1000 + 100 - 662 - 50)}{1462 - 718} = 0.141.$$

The work produced by the steam turbine per unit of gas turbine mass flow is $(h_c - h_{d'})(\dot{m}_{H_2O}/\dot{m}_{gas})$.

In the gas cycle we note that for a fixed T_{t4} the turbine temperature ratio is known, since $T_{t5} = T_c + \Delta T_s$ is fixed by the steam cycle, that is, $\tau_t = T_{t5}/T_{t4} = (T_c + \Delta T_s)/T_{t4}$. The turbine pressure ratio follows from $\pi_t = [1 - (1 - \tau_t)/\eta_t]^{\gamma_t/(\gamma_t - 1)}$. Tracing the stagnation pressure ratios through the engine

$$\pi_c\pi_b\pi_t = 1 \tag{3.22}$$

where π_b represents the pressure ratio of the combustor and boiler together. Thus π_c is set, and

$$\tau_c = 1 + \{[\pi_c^{(\gamma_c - 1)/\gamma_c} - 1]/\eta_c\}.$$

The work of the gas turbine is then

$$\dot{m}_{gas}[c_{pt}T_{t4}(1 - \tau_t) - c_{pc}T_0(\tau_c - 1)]$$

while the heat input to the combustor is $\bar{c}_p(T_{t4} - T_0\tau_c)$. Finally, the efficiency of the combined cycle is the sum of gas turbine and steam turbine works divided by the heat input, or

$$\eta_{combined} = \frac{c_{pt}}{\bar{c}_p}\left[1 - \frac{T_c + \Delta T_s}{T_0\theta_t} + \left(\frac{T_c + \Delta T_s}{T_0} - \frac{T_b + \Delta T_b}{T_0}\right)\left(\frac{h_c - h_{d'}}{h_c - h_b}\right)\right.$$
$$\left. - \left(\frac{\pi_c^{(\gamma_c - 1)/\gamma_c} - 1}{h_c}\right)\left(\frac{c_{pc}}{c_{pt}}\right)\right]\bigg/\left[\theta_t - 1 - \frac{\pi_c^{(\gamma_c - 1)/\gamma_c} - 1}{\eta_c}\right].$$

Continuing the above example, we find for $\theta_t = 5$, $T_0 = 500°R$, $\tau_t = 1560/2500 = .624$, and $\pi_t = 0.1002$ if $\eta_t = 0.85$. Then for $\pi_b = 0.95$, $\pi_c = 10.5$, and from (3.23), $\eta = 0.529$.

The results of a series of such calculations, for different steam pressures and maximum steam temperatures, are summarized in figure 3.8 where the lowest curve is for saturated steam (no superheat) and the upper two are for different superheat temperatures. The highest efficiencies are attained with superheated steam cycles since they provide a closer match between the heat

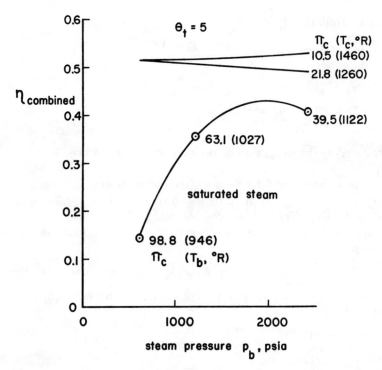

3.8 Efficiency of the gas turbine-steam combined cycle as a function of steam pressure for superheated (upper curves) and saturated (lower curve) steam cycles.

rejection curve of the gas turbine and the heat absorption curve of the steam. The efficiency is insensitive to steam pressure, with a slight advantage for the higher pressures. Note that peak efficiencies near 0.52 are predicted for this cycle.

Using π_c as a parameter, the efficiency of the gas turbine–steam cycle is compared to that of the regenerative gas turbine in figure 3.7. It offers about a 0.15 improvement, a large margin for ground-based power plants.

3.10 Concluding Comments

Variations on the theme of this chapter could fill a much larger book than this one. The possible range of parametric variations even for the simple turbojet is such that there is little sense in attempting to produce a complete set of characteristics. Rather, the characteristics of a particular engine cycle must be calculated when the need arises. If these examples aid the reader in this process, they will have served their purpose.

A great many variations of the basic turbojet, turbofan, turboshaft, and other cycles can be treated by slight modifications of this approach. These include turbine cooling and compressor bleed. Suggestions for such analyses will be found in the exercises at the end of this chapter, and some of their ramifications will be discussed in later chapters.

References

3.1 J. H. Keenan and F. G. Keyes, *Thermodynamic Properties of Steam*, John Wiley, 1936.

Problems

3.1 Write expressions for $F/\dot{m}a_0$ and I for a ramjet; include all appropriate losses π_d, π_b, and so on, and account for changes in c_p with temperature. Assume the nozzle is ideally expanded.

3.2 Write expressions for $F/\dot{m}(1 + \alpha)$ and for I for a dual-nozzle turbofan (figure 1.6), including the effects of inefficiency in the compression and expansion processes and of changes in c_p with temperature. Assume that both nozzles are ideally expanded and that the jet exit velocities are equal. (Do not try to write this last condition out explicitly; a numerical iteration is easier.)

3.3 Derive expressions for $F/\dot{m}a_0$ and I for a turbojet that is ideal except that it has a choked convergent nozzle so that $M_6 = 1$. Compare your results numerically to those of figure 2.3.

4
Nonrotating Components

In chapters 2 and 3, engine performance was related to the performance of the several major components of the engine—inlet, compressor, burner, turbine, and nozzle; each component was described by the parameters through which it influences the thermodynamic cycle. Chapters 4 through 8 show how the behavior of each of the components is determined by its mechanical characteristics and what factors limit the performance of the components, and hence of the engine.

Some of these limiting factors stem from fluid mechanical phenomena, some from thermal effects, and some from mechanical or structural limitations. Even if they could all be quantified, a complete, quantitative treatment of all of them would require many volumes, and no such treatment will be attempted here. Rather, the phenomena will be discussed in physical terms, with as much analysis as is required to illustrate the important compromises controlling the design of any engine.

Nevertheless, some background is required in gas dynamics and solid mechanics. Particularly essential is an understanding of compressible channel flow, shock waves, and the rudiments of boundary layers. Since these phenomena will play a central role in much of the discussion to follow, a brief summary of each will be given. The reader with no prior exposure to compressible flow would do well to study the relevent sections of one of the excellent texts, such as reference 4.1 or 4.2.

4.1 Summary of Gas Dynamics

Three topics from gas dynamics will be discussed here; channel flow, because it provides an intuitive understanding of the way flow through the internal passages of engines is controlled by the shape of the passage walls; shock waves, because they lead to large qualitative differences between subsonic and supersonic flows; and boundary layers, because they lead to important limitations on the performance of the engine components.

4.1.1 Channel Flow
By "channel flow" we mean flow through a closed passage of gently varying cross-sectional area, such that the velocity component along the axis of the channel is much larger than the components perpendicular to the axis. When this condition is met, the kinetic energy of the flow and its momentum may be assumed to be due to only the one velocity component. In this sense the flow is "one dimensional," but the variations in pressure, velocity, temperature, and so forth, all result from the small velocities, perpendicular to the axis, which result from the shape of the channel wall. These effects are brought into the analysis by the continuity equation in the form

$$\rho u A = \dot{m} = \text{const.,} \tag{4.1}$$

where A is the cross-sectional flow area of the duct. Combining the first law of thermodynamics and the axial momentum equation gives an equation for the total energy, or stagnation enthalpy h_t, in terms of the thermal enthalpy and kinetic energy,

$$h + u^2/2 = h_t = \text{const.,} \tag{4.2}$$

where the subscript t denotes the stagnation state. If the flow is isentropic, the entropy is constant, or

$$s = s_t = \text{const.} \tag{4.3}$$

If the gas is thermally perfect,

$$p = \rho R T. \tag{4.4}$$

In most of the following discussion the gas will also be assumed to be calorically perfect so that $h = c_p T$ where c_p is constant. Then from (4.2),

$$c_p T(1 + u^2)/2c_p T = c_p T\{1 + \tfrac{1}{2}(\gamma - 1)M^2\} = c_p T_t, \quad \text{or}$$
$$T_t/T = 1 + \tfrac{1}{2}(\gamma - 1)M^2. \tag{4.5}$$

Since $s = s_t$ and $s = s_t + c_p \log(T/T_t) - R \log(p/p_t)$, we have from (4.3)

$$p_t/p = (T_t/T)^{\gamma/(\gamma-1)} = \{1 + \tfrac{1}{2}(\gamma - 1)M^2\}^{\gamma/(\gamma-1)}. \tag{4.6}$$

The mass flow per unit area is $\rho u = (p/RT)M(\gamma RT)^{1/2}$ so that

$$\rho u/\rho_t a_t = M/\{1 + \tfrac{1}{2}(\gamma - 1)M^2\}^{(\gamma+1)/2(\gamma-1)} \tag{4.7}$$

where a_t is the speed of sound at stagnation conditions. For given ρ_t and a_t, ρu is a maximum for $M = 1$. This condition is denoted by an asterisk:

$$(\rho u)^*/\rho_t a_t = [2/(\gamma + 1)]^{(\gamma+1)/2(\gamma-1)} \tag{4.8}$$

Then from (4.1),

$$\frac{A}{A^*} = \frac{(\rho u)^*}{\rho u} = \frac{1}{M}\left(\frac{1 + \tfrac{1}{2}(\gamma - 1)M^2}{\tfrac{1}{2}(\gamma + 1)}\right)^{(\gamma+1)/2(\gamma-1)}. \tag{4.9}$$

This relation connects the various flow properties to A. It is plotted in figure 4.1. For the present, we simply note that ρu peaks at $M = 1$ and drops off by about a factor of 5 from this peak value at $M = 3$. This large range of variation greatly complicates the design of supersonic inlets. From (4.8),

$$(\rho u)^* = \rho_t a_t [2/(\gamma + 1)]^{(\gamma+1)/2(\gamma-1)}$$
$$= \gamma^{1/2}[2/(\gamma + 1)]^{(\gamma+1)/2(\gamma-1)} p_t/(RT_t)^{1/2}. \tag{4.10}$$

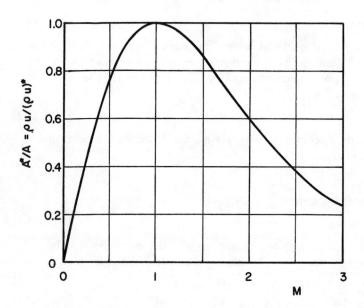

4.1   Ratio of mass flow density to that at $M = 1$ as a function of M for isentropic channel flow.

For air at $p_t = 1$ atm and $T_t = 288°K$, $(\rho u)^* = 240.7 \text{ kg/m}^2\text{s}$. It increases directly with p_t and decreases as the square root of T_t. These variations of $(\rho u)^*$ with p_t and T_t are keys to the behavior of the flow in the engine.

4.1.2   Shock Waves

By changing the entropy and hence the stagnation pressure of the bulk flow away from solid surfaces, shock waves exert a strong influence on the behavior of supersonic flows. Their existence is explained by the fact that the equations of conservation of momentum, mass, and energy have not a unique solution but rather two solutions, one corresponding to $M > 1$ and one to $M < 1$. Writing these relations across a discontinuity we have

$$\rho_1 u_1 = \rho_2 u_2,$$

$$p_1 + \rho_1 u_1^2 = p_2 + \rho_2 u_2^2,$$

$$c_p T_1 + u_1^2/2 = c_p T_2 + u_2^2/2,$$

where u is the velocity *perpendicular to the discontinuity* and these can be manipulated to give

$$u_1 u_2 = (a^*)^2,$$

$$M_2^2 = \{1 + \tfrac{1}{2}(\gamma - 1)M_1^2\}/\{\gamma M_1^2 - \tfrac{1}{2}(\gamma - 1)\},$$

$$p_2/p_1 = [2\gamma/(\gamma + 1)]M_1^2 - [(\gamma - 1)/(\gamma + 1)],$$

$$T_{t2} = T_{t1},$$

and

$$p_{t1}/p_{t2} = \{[2\gamma M_1^2/(\gamma + 1)] - [(\gamma - 1)/(\gamma + 1)]\}^{1/(\gamma - 1)}$$

$$\cdot \{[(\gamma - 1)M_1^2 + 2]/(\gamma + 1)M_1^2\}^{\gamma/(\gamma - 1)} \tag{4.11}$$

This last equation, giving the stagnation pressure ratio across the shock as a function of the normal upstream Mach number, is particularly important for the following arguments. It is plotted on figure 4.2 with p_1/p_2, the static pressure ratio across the shock. Note that for M_1 slightly larger than unity, the shock produces a significant static pressure ratio with little stagnation pressure loss, but as the Mach number increases above about 1.5, the loss increases rapidly.

Since $(\rho u)^* \propto p_t$ and T_t is constant across the shock,

$$(\rho u)_2^*/(\rho u)_1^* = p_{t2}/p_{t1}. \tag{4.12}$$

These relations apply equally to all shocks, but the convention is to refer to the shock as "normal" if it stands perpendicular to the total flow velocity, so that M_1 is the Mach number of that total velocity. If the shock stands at

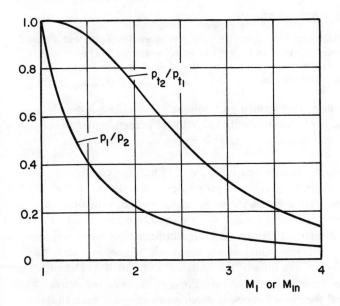

4.2   Stagnation pressure ratio and static pressure ratio across shock as function of normal upstream Mach number.

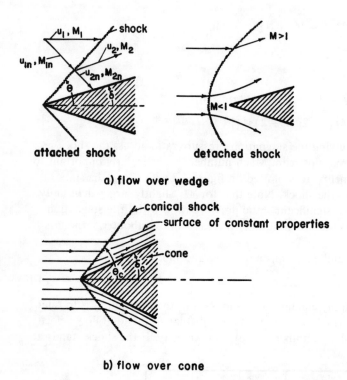

a) flow over wedge

b) flow over cone

4.3   a. Supersonic flow over two-dimensional wedge with attached flow at left and detached flow at right, showing definitions of wave angle θ and deflection (wedge half) angle δ.
b. Supersonic flow over axisymmetric cone, showing conical shock, and continued compression of flow over surface of cone.

an angle to the total flow, it is termed an "oblique" shock, and $M_1 = M_{1n}$ is then the Mach number associated with the velocity perpendicular to the shock surface. Such an oblique shock can be generated by a wedge, as shown in figure 4.3a.

The shock is completely determined by M_1 and θ, but it is convenient to relate the wave angle θ, and therefore all properties of the shock, to the deflection angle δ since this specifies the shape of the wedge. When this is done, the results shown in figures 4.4 and 4.5 are found (reference 4.3). For a given δ, there is a minimum M_1 below which the oblique shock will not turn the flow sufficiently. For lower values of M_1 the shock "detaches," forming a normal shock in front of the nose of the wedge as in figure 4.3. The subsonic flow behind this normal shock can turn through the required deflection angle. The flow behind the curved shock is much more complex than that behind a simple oblique shock because it is partially subsonic and partially supersonic.

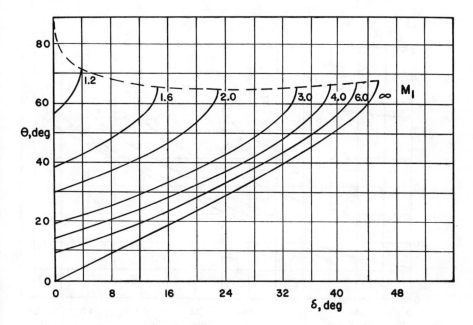

4.4   Wave angle θ as a function of deflection angle δ and initial Mach number M_1 for oblique shock on two-dimensional wedge.

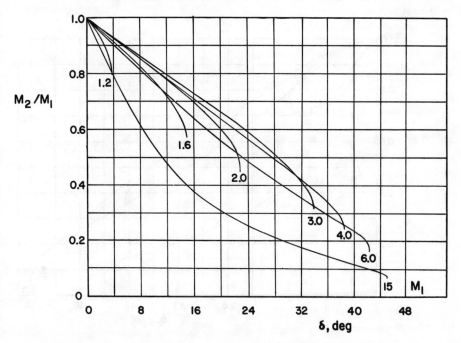

4.5   Ratio of final to initial Mach number M_2/M_1 as a function of deflection angle δ and initial Mach number M_1 for oblique shock on two-dimensional wedge.

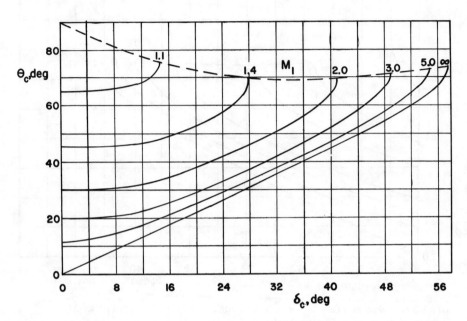

4.6 Wave angle as a function of cone half angle and initial Mach number for flow over axisymmetric cone.

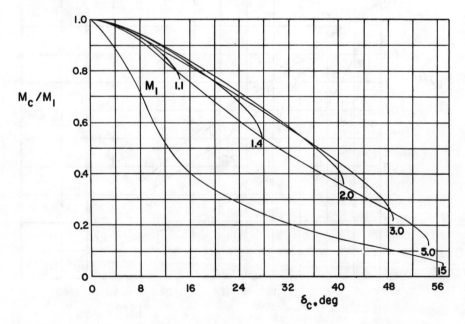

4.7 Cone surface Mach number M_c as a function of cone half angle δ_c and initial Mach number M_1 for flow over axisymmetric cone.

For the two-dimensional wedge with attached shock, the flow behind the shock is completely uniform and parallel to the surface of the wedge. The analogous situation for an axisymmetric cone is that the flow has constant properties along lines radiating from the tip of the cone. Because the shock surface is made up of such lines, it has constant normal Mach number, pressure ratio, and stagnation pressure loss. The surface of the cone is another surface of constant properties, but the flow field outside the cone has curved streamlines, as sketched in figure 4.3b. The variation of shock angle and Mach number at the cone surface are given for such flows in figures 4.6 and 4.7 (reference 4.3), because they are useful for estimating the performance of axisymmetric inlets.

4.1.3 Boundary Layers and Separation

Blading and passages are designed to control the air movement in an engine to produce the directions and velocities required for energy exchange with the fluid, or between its thermal and kinetic energies, with the minimum entropy production (losses). In the main the fluid is controlled by pressure gradients, which are induced by pressure forces acting on the surfaces controlling the flow. Thus, in the simple divergent passage sketched in figure 4.8, the (subsonic) flow is decelerated (diffused) primarily by pressure forces acting on the divergent walls. In the main part of the flow viscous effects are small because the shears are small.

But in the immediate neighborhood of the walls, the velocity must change rapidly from the bulk value to zero to satisfy the no-slip condition at the surface. As a result, in a region near the wall the behavior of the fluid is controlled by viscous shear forces. If this region is very thin compared to the dimensions of the channel (or in the case of a blade, compared to its chord), the overall flow may be well represented by that in a passage made slightly smaller to allow for the viscous region (boundary layer) but of essentially the "inviscid" shape. One of the purposes of boundary layer theory is to define the correction to passage shape, termed the "displacement thickness" of the

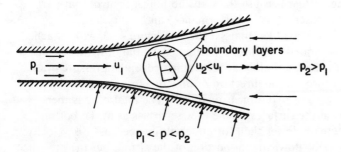

4.8 Diffusion in a subsonic channel flow.

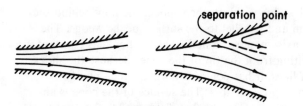

4.9   Unseparated (left) and separated flow in a subsonic channel.

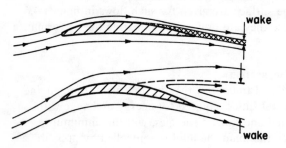

4.10   Unseparated (top) and separated flow (bottom) on the suction surface of a blade, showing increase in wake width.

boundary layer. A more difficult and more important task of the theory is to determine when the flow will fail to follow the surfaces of the passage because of the presence of the viscous region. This failure is termed "separation."

To appreciate the importance of separation, we must recognize that for almost all gasdynamic devices, the viscous shear forces acting on the surfaces are small compared to the pressure forces, which largely govern the behavior of the flow. Thus in the channel in figure 4.9, if the flow remains attached to the walls as in the left sketch, the loss associated with viscous effects is of the order of the ratio of boundary layer thickness to channel dimension—ordinarily a very small number. On the other hand, if the flow separates from the wall as in the right sketch, the diffusing effect of the downstream portion of the passage is lost because the walls no longer control that part of the flow, and the diffuser's loss in performance can be very large.

Another example is a blade like that sketched in figure 4.10, with attached flow at the top and separated flow at the bottom. Potential flow theory tells us that if the streamlines close smoothly at the trailing edge, the pressure drag of the airfoil (the net force in the direction of flow due to normal forces on the airfoil surface) is zero in two-dimensional flow. The total drag is then due to viscous shear on the surfaces. If the flow separates as at the bottom, the viscous drag is actually lower than for attached flow because the velocity gradient is reduced on the downstream portion of the airfoil, but the pressure drag increases greatly because the pressure over the downstream portion of the suction side of the airfoil is lower than it would be if the flow were smoothly diffused as in the attached case. The increased drag generates a broader wake of low-momentum fluid than for the case of attached flow, and the mixing of this wake with the inviscid outer flow ultimately leads to an entropy increase in the flow.

Boundary layer theory is one of the more difficult branches of fluid mechanics, and the prediction of separation is one of the most difficult aspects of boundary layer theory; thus no quantitative treatment of this controlling phenomenon can be attempted within the scope of this book. The discussion will be limited to qualitative descriptions of the important phenomena. For quantitative treatments, see reference 4.4.

Consider the region of viscous flow near the wall, as sketched in figure 4.11. The flow velocity along the surface u increases from zero at the wall to the free-stream value at the edge of the viscous region. The viscous shear force at the wall is transferred outward by shear forces in the fluid, and decelerates the flow in the viscous region. In the absence of pressure gradients along the flow direction, this deceleration causes the thickness of the viscous region to grow in the flow direction; that is, the boundary layer thickens as it entrains more fluid from the inviscid flow.

If there is a pressure gradient in the flow direction, the rate of change of the boundary layer thickness is altered. The reason is that to give any particular pressure change, a larger fractional change of the flow velocity is required in the boundary layer than is required in the inviscid flow, because the velocity is lower in the boundary layer. Thus if the free-stream pressure decreases along the flow, the boundary layer fluid is *accelerated* relative to the inviscid flow, the boundary layer is thinned, and we say the pressure gradient is favorable. If the pressure increases in the flow direction, the boundary layer fluid is *decelerated* relative to the inviscid flow, and the boundary layer is thickened. The greater thickness implies *smaller* viscous shear than in the case of zero-pressure gradient, and if the pressure gradient

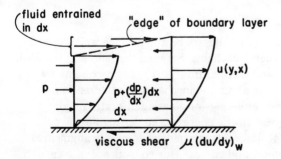

4.11   The forces acting on an element of boundary layer flow.

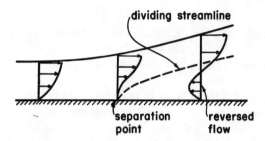

4.12   Flow at a two-dimensional separation point in a boundary layer.

is too large, it will overcome the shear forces that transfer momentum toward the wall; thus the fluid near the wall will stagnate and, in fact, reverse, as sketched in figure 4.12. This is what is meant by "separation" in two-dimensional flow.

Laminar boundary layers can support only a very small amount of diffusion prior to separation—on the order of a 10 percent reduction in flow velocity if the velocity decreases linearly. Turbulent boundary layers fortunately do better, and one method of preventing separation is to induce transition to turbulent flow. Most aircraft engine components operate at Reynolds numbers such that transition to turbulent flow occurs on the blades. Measurements in straight-walled, two-dimensional diffusers have shown incipient separation for wall divergence half-angles larger than about 5 degrees.

Although much effort has been devoted to developing methods for predicting separation of turbulent boundary layers, no methods are available at present to predict separation on the blades or casings of turbomachines. Unsteadiness and three-dimensional effects make the situation so complex that separation is even hard to define. The most fruitful approach has been to use some of these ideas to correlate cascade performance data.

Shock waves can induce severe separation problems when they impinge on walls by imposing sudden pressure changes on the boundary layers.

4.1.4   Gas-Solid Heat Transfer
Wherever the fluid exerts viscous shear stresses on the passage walls, there is also the possibility of thermal energy exchange between the fluid and the wall. In gases, momentum transport and thermal transport occur by essentially the same mechanisms. For laminar flows, the mechanism is the random motion of molecules, which leads molecules originating in hotter regions of the fluid to transfer a part of their kinetic energy to cooler regions by collisional interaction. For turbulent flows this mechanism is greatly augmented by more-or-less random motion of fluid eddies. This equivalence of transport

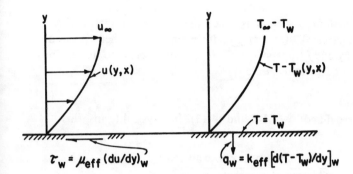

4.13 Illustrating the analogy between momentum transport in a velocity gradient (left) and heat transfer in a temperature gradient (right).

mechanisms does not exist for all fluids. In liquid metals, for example, electronic transport is a major factor in thermal conduction.

Restricting ourselves to ordinary gases, we can construct an analogy between heat transfer and shear along the lines indicated in figure 4.13. The shear stress at the wall τ_W is related to the velocity gradient by an effective viscosity μ_{eff} equal to the fluid viscosity for laminar flow and larger for turbulent flow. For low Mach numbers, the driving force for heat transport to the wall is the difference between fluid and wall temperatures; thus if the momentum and energy are transported by the same mechanism, we would expect the variations of $u(y)$ and $T(y) - T_W$ to be similar. To make this statement quantitatively, we must nondimensionalize the shear and heat transfer. Customarily, we nondimensionalize the shear by the free-stream dynamic pressure to yield a friction coefficient

$$c_f \equiv 2\tau_W/(\rho u_\infty^2) = 2\mu_{eff}(\partial u/\partial y)_W/\rho_\infty u_\infty^2. \tag{4.13}$$

Similarly, we divide the heat flux by the rate at which thermal energy (referenced to the surface temperature) is convected along the surface to yield a Stanton number

$$St \equiv \frac{q_W}{\rho_\infty u_\infty c_p(T_\infty - T_W)} = \frac{k_{eff}\partial(T - T_W)/\partial y|_W}{\rho_\infty u_\infty c_p(T_\infty - T_W)}. \tag{4.14}$$

Now the statement that the variations of $u(y)$ and $T(y) - T_W$ should be similar implies that

$$[1/(T_\infty - T_W)]\partial(T - T_W)/\partial y|_W = (1/u_\infty)(\partial u/\partial y)|_W,$$

so we see that this is equivalent to

$$St = [k_{eff}/c_p\mu_{eff}](c_f/2).$$

The dimensionless group $c_p\mu/k = Pr$ is called the Prandtl number, and is near unity for most gases (0.7 for air). For turbulent flows, where k_{eff} and μ_{eff} are dominated by eddy transport, $c_p\mu_{\text{eff}}/k_{\text{eff}}$ is quite close to unity. Thus we arrive at a form of Reynolds analogy,

$$St \approx c_f/2, \tag{4.15}$$

where St and c_f are defined by (4.14) and (4.13). In physical terms, this relation states that

$$\frac{\text{Heat flux to wall}}{\text{Convected heat flux}} = \frac{\text{Momentum flux to wall}}{\text{Convected momentum flux}}$$

or

$$q_W/[(\rho_\infty u_\infty)c_p(T_\infty - T_W)] = \tau_W/[(\rho_\infty u_\infty)u_\infty].$$

We will use it to establish some important characteristics of cooled turbines in section 6.3.

4.2 Diffusers

The function of the diffuser (or inlet) is to bring the air from ambient conditions to the conditions required at inlet to the engine. As noted in section 1.12 and indicated schematically in figures 1.13 and 1.14, the character of the diffuser is very different for purely subsonic flight than for supersonic flight where the requirements for decelerating the flow from supersonic speeds are added to those the subsonic diffuser must meet.

In this section we must refer to the mass flow characteristics of compressors that have not been described yet. If these remarks are less than satisfying, refer to chapter 5 where these characteristics are discussed. Because of the close interrelationships between the engine components, it is not possible to avoid totally such references.

4.2.1 Subsonic Diffusers
Two main constraints must be met by the subsonic diffuser. It must supply air to the engine at the axial Mach number M_2 demanded by it. As we shall see, M_2 is mainly determined by the rotational Mach number of the compressor or fan. It therefore depends on rotational speed and inlet air temperature and is largest at conditions of high altitude (low T_0) and full engine speed. M_2 is smallest for conditions of low altitude (high T_0) and low engine speed. For a typical subsonic transport application the most important requirements are for takeoff at full engine speed and high T_0, and for cruise at lower T_0 (but at $M_0 \approx 0.8$) and reduced engine speed. The decrease in T_0 tends to be offset by the increase in M_0 and the reduction in engine speed,

so the variation in M_2 is not large. A decrease of about 20 percent from takeoff to high subsonic cruise is typical.

A more serious problem is the change in M_0 from zero at takeoff to about 0.8 at cruise. With the M_2 nearly fixed this imposes large changes in the geometry of the streamlines entering the diffuser with attendant problems of boundary layer separation when the inlet surfaces must turn the flow through large angles. From the curve of mass flow density ρu versus M of figure 4.1, the ratio of inlet streamtube area to engine inlet area A_0/A_2 can be computed; it is $\rho u(M_2)/\rho u(M_0) = [A^*/A(M_2)]/[A^*/A(M_0)]$. If $M_2 = 0.5$, then A_0/A_2 varies from ∞ at $M_0 = 0$ to 0.78 at $M_0 = 0.8$. The shape of the captured streamtube therefore varies somewhat as shown in figure 4.14, and the problem is to design an inlet that will accept the flow with the indicated large changes in direction without excessive losses due to separation.

If optimized for the $M_0 = 0.8$ cruise, the inlet would have the dashed contour in the lower half of figure 4.14 with a thin lip to minimize the increase in Mach number as the flow is divided. But this inlet would separate badly on the inside at takeoff and low subsonic conditions because the turn around the sharp lip would impose severe pressure gradients. The compromise is to round the lip, making it less sensitive to flow angle, but incurring some loss due to separation in the exterior flow.

In addition, "blow-in doors" are sometimes added to allow some flow to bypass the minimum area of the inlet and so reduce separation problems at takeoff. These doors are open only at full-throttle conditions during takeoff. They were used in the turbofan-powered B-707, and in the early model of the B-747. It is interesting to note that the blow-in doors have been deleted

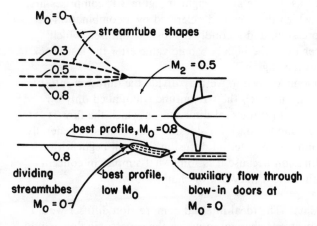

4.14 Schematic of a subsonic diffuser; the top shows the streamtube shapes for various flight Mach numbers, and the bottom shows the compromises in design necessitated by them.

from later model B-747s, because the nonuniform flow into the fan, which resulted from their opening on takeoff, led to unacceptable increases in fan noise. Eliminating the doors required an increase in minimum area, a more rounded lip, and presumably somewhat more drag at cruise conditions.

The details of the design of a subsonic inlet depend on the way these compromises are struck. Calculations of the boundary layer flows over the inner and outer surfaces can serve as a guide during the design process, but some empirical development is also generally required. When fully developed, a good inlet will produce a pressure recovery π_d between 0.95 and 0.97 at its optimum condition.

4.2.2 Supersonic Diffusers

Flight at supersonic speeds complicates the diffuser design for three reasons. The most fundamental is the existence of shock waves, which introduce a wholly new loss mechanism that can lead to large decreases in stagnation pressure even in the absence of viscous effects and to bistable operation, with very large losses in one mode and much smaller losses in the other. Much of the emphasis in discussions of supersonic diffusers has been on this aspect of the problem. A second reason is that the variations in capture streamtube diameter between subsonic and supersonic flight for a given engine are very large, as much as a factor of four between $M_0 = 1.0$ and $M_0 = 3.0$, and the aircraft that are to fly at $M_0 = 3.0$ must also operate at $M_0 = 1$! Finally, as M_0 increases the inlet compression becomes a larger fraction of the overall cycle compression ratio; as a result the specific impulse and thrust per unit of mass flow become more sensitive to diffuser pressure recovery.

A typical diffuser, such as the one at the right in figure 3.1, comprises first a supersonic diffuser, in which the flow is decelerated by a combination of shocks and diffuse compression, and secondly a subsonic diffuser, which reduces the Mach number from the high subsonic value after the last shock to the value acceptable to the engine.

Focusing on the supersonic diffusion, we may divide the diffusion into two types: (1) external diffusion, in which the streamtube is bounded only on one side by solid surfaces, and (2) internal diffusion, in which the flow is through a channel, hence bounded on both sides. Modern diffusers generally use external diffusion, followed by internal diffusion. We will begin with a description of internal diffusion because it is somewhat simpler in concept and can be discussed within the framework of channel flow.

4.2.2.1 Internal Compression

The ideal internal compression diffuser would consist of a convergent-divergent channel in which the supersonic flow would be decelerated by a diffuse series of weak compression waves to sonic velocity, then diffused subsonically. While such a diffuser could work in principle, it is impractical for at least two reasons. One is the starting problem. The ideal shock-free diffusion cannot be attained by increasing the Mach number M_0 from below to the final value with fixed geometry of the diffuser. The second is that boundary layer growth prevents diffusion to $M = 1$. Recall (figure 4.1) that the passage area passes through a minimum at $M = 1$; thus for a given Mach number at entrance to the diffuser, the Mach number at the point of minimum area (throat) would be very sensitive to the thickness of the boundary layer there. As a result the flow would shock, and the shock would be expelled to a position ahead of the diffuser entrance.

Addressing first the starting problem, we consider the development of the flow with increasing flight Mach number for a diffuser of area ratio, A_c/A_t capture to throat, as shown in figure 4.15. The ideal, shock-free operation (figure 4.15a) would require a value of M_0, such that $A(M_0)/A^* = A_c/A_t$. For any smaller M_0, $A(M_0)/A^* < A_c/A_t$, and if $A^* = A_t$, then $A(M_0) < A_c$; the streamtube that can be captured is smaller than A_c. The excess mass flow must be diverted around the inlet; but this cannot occur if the flow is supersonic all the way to the inlet lip, so a shock forms ahead of the lip and the

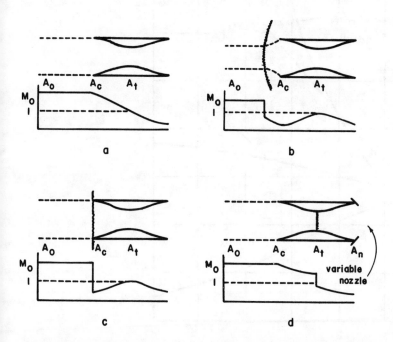

4.15 Schematics of internal compression diffuser, showing (a) ideal isentropic diffusion from M_0 through unity to $M < 1$, (b) operation below the critical (starting) Mach number, (c) at the critical $M_{0,c}$ but not started, and (d) at the critical $M_{0,c}$, and started, with shock positioned at throat.

spill of excess mass flow occurs in the subsonic flow behind it, as in figure 4.15b. To determine what fraction of the mass flow will be spilled at any M_0 and A_c/A_t, we must take account of the decrease in stagnation pressure, hence in $(\rho u)^*$, across the shock. The ratio $(\rho u)_1^*/(\rho u)_0^* = p_{t1}/p_{t0}$ across the shock is given by figure 4.2 or equation (4.11) as a function of M_0. Now equating the flow through the throat to the flow in the captured streamtube gives $\rho_0 u_0 A_0 = (\rho u)_0^* A_t (p_{t1}/p_{t0})$ or,

$$A_0/A_t = [A(M_0)/A^*][p_{t1}(M_0)/p_{t0}]. \qquad (4.16)$$

This ratio of actual captured streamtube area to throat area is shown in figure 4.16 as a function of M_0, along with the corresponding ratio (simply A/A^*) for the shock-free diffuser. When M_0 reaches the critical value M_{0c} such that $A_0/A_t = A_c/A_t$, the normal shock will stand just at the lip, as in figure 4.15c. But in this position it is unstable and will move downstream if perturbed.

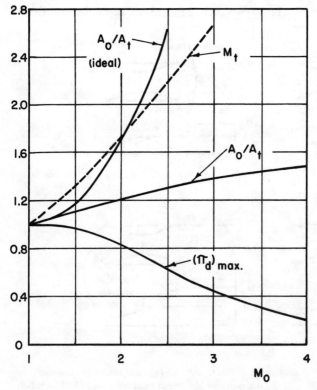

4.16 Characteristics of a fixed-geometry internal compression inlet, including best pressure recovery, capture-to-throat area ratio, and throat Mach number.

This is easily seen by imagining that the shock is moved slightly downstream by some disturbance. As the Mach number decreases downstream with supersonic flow in the converging passage, the perturbed shock will stand at a lower Mach number and will cause less stagnation pressure loss. The throat remaining choked will then pass more mass flow than is captured by the lip with a consequent net outflow of mass from the convergent section of the diffuser. This requires a reduction in average density that can only be affected by discharging the shock downstream, thus "starting" the diffuser.

To achieve the best possible pressure recovery with the diffuser, once started, the back pressure is adjusted so that the shock stands at the throat, where the Mach number is smallest, as in figure 4.15d.

The Mach number M_t ahead of the shock at the throat is readily found by noting that it corresponds to a value of A/A^* given by

$$A(M_t)/A^* = [A(M_0)/A^*][A_t(M_0)/A_0].\qquad(4.17)$$

The best pressure recovery this inlet can achieve is that for a normal shock at M_t.

Values of M_t and $(\pi_d)_{max}$ for such inlets are given in figure 4.16. The pressure recovery is good up to $M_0 \approx 1.5$ but drops off rapidly at higher Mach numbers.

4.2.2.2 External Compression Much better pressure recovery can be had at high M_0 by taking advantage of a characteristic of shock waves: for a given overall pressure (or Mach number) ratio, a series of weak shocks produces much less stagnation pressure loss than one strong shock. The simplest use of this advantage is in the single-oblique-shock diffuser; a two-dimensional version of it is shown in figure 4.17. The oblique shock decelerates the flow to some M_1 between M_0 and 1, and it is again decelerated through a nearly normal shock at the throat.

To illustrate the characteristics of such a diffuser, consider a design for $M_0 = 4$. Taking a set of three deflection angles δ, one finds the shock angles θ from figure 4.4 and M_1 from figure 4.5. The stagnation pressure ratio

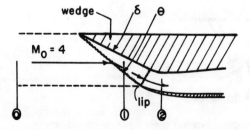

4.17 Schematic of single-oblique-shock external compression inlet designed for $M_0 = 4$.

across the oblique shock follows from figure 4.2 with $M_{1n} = M_0 \sin \theta$, and that across the normal shock directly from figure 4.2 with $M_{1n} = M_1$.

$\delta =$	20°	25°	30°
$\theta =$	32°	38°	45°
$M_1 =$	2.57	2.21	1.85
$p_{t1}/p_{t0} =$	0.67	0.51	0.37
$p_{t2}/p_{t1} =$	0.47	0.62	0.79
$\pi_d =$	0.315	0.316	0.295

Evidently the best wedge angle for $M_0 = 4$ is about 22.5 degrees. Comparing the pressure recovery of 0.32 to that for a simple internal compression inlet, 0.2 (see figure 4.16), we see a marked improvement.

Further gains can be made by introducing more oblique shocks. Three, as at the top of figure 4.18, would give $\pi_d = 0.63$ at $M_0 = 4$. The limit in this progression is the isentropic wedge diffuser at the bottom of figure 4.18 in which compression takes place to $M = 1$ through a series of very weak compression waves. There are some serious difficulties with such high performance diffusers, however. One is that the shocks originating on the wedge in the multiwedge diffuser will tend to separate the boundary layer there. The streamwise pressure increase along the spike has the same effect in the isentropic diffuser. Another disadvantage is that, as the compression approaches isentropy, the flow deflection angle increases from zero for a normal shock to 68 degrees at $M_0 = 4$ for an isentropic diffuser. The near-sonic flow at this angle must be turned back to the axial direction, and this leads to a

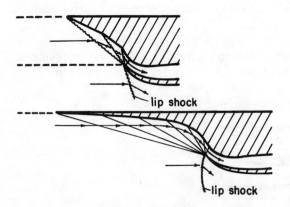

4.18 Schematics of (upper) three-oblique-shock external compression inlet and (lower) isentropic spike inlet, showing how air turning angle increases with approach to isentropy.

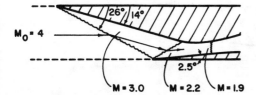

4.19 Mixed-compression inlet configured for $M_0 = 4$.

large angle on the lip which will form a shock and produce considerable
"cowl drag." It is clear that a compromise must be made between pressure
recovery and drag.

4.2.2.3 Mixed Compression   By combining an initial external compression
through an oblique shock with internal compression inside the lip, one can
approach the high-pressure recovery of the oblique-shock diffuser without
incurring as much cowl drag as with a pure external compression inlet. The
scheme is indicated in two dimensions in figure 4.19. The three-shock diffuser
shown would have an ideal $\pi_d = 0.55$ at $M_0 = 4$. This is considerably better
than the 0.32 of the two-shock external compression inlet of figure 4.17, and
the diffuser as shown would have low cowl drag. The main disadvantage is
that the internal contraction reintroduces the starting problem, but this difficulty can be overcome by reasonable geometry variations in flight. An inlet
similar to that of figure 4.19 was under development for the B2707–300 (SST).

4.2.2.4 Axisymmetric Diffusers   Thus far the discussion of supersonic diffusers
has dealt with two-dimensional devices. Although many practical inlets are
essentially two-dimensional, in some applications an axisymmetric inlet is
desirable, as in pod-mounted engine installations. Indeed, the popular con-
ception of a supersonic inlet is axisymmetric.
    The flow in such an inlet is more complex than in two-dimensional ones
because the shocks are curved rather than plane surfaces and in general are
of varying strength over their surface. Fortunately some simplification results
from the characteristics of conical flow fields (section 4.1.2). The shock generated by a cone, as indicated in the cross-sectional schematic of figure 4.20,
has constant properties, pressure ratio, and turning, but the flow behind it
undergoes a continued isentropic compression. This is an advantage because
the flow along the cone surface is compressed isentropically, but it is a disadvantage because the flow at the lip, just behind the shock, has a higher
Mach number than that at the cone surface. In fact, it is the same as for a
two-dimensional inlet with the same shock angle. Both effects are rather
small in practice.

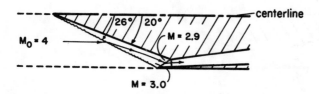

4.20 Axisymmetric mixed-compression inlet configured for $M_0 = 4$; compare Figure 4.18.

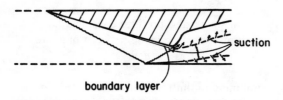

4.21 Boundary layer control in a mixed-compression inlet, showing suction slot at shock impingement point on wedge and porous suction in subsonic diffuser.

A direct comparison of the two types of inlet can be made by choosing the cone angle so the initial oblique shock is the same strength (lies at the same angle) as in the two-dimensional diffuser. This has been done in figure 4.20 as compared to figure 4.19; the cone angle and surface Mach number are taken from figures 4.6 and 4.7. Note that the cone surface Mach number differs only slightly from that at the lip, so that the second shock would be of nearly uniform strength in this example, and the pressure recovery would be almost the same as for the two-dimensional inlet.

4.2.2.5 Boundary Layer Control   Each of these external compression inlets has at least one point where a shock impinges on a surface. As noted in section 4.1.3, this is almost sure to result in separation of the boundary layer unless some preventive action is taken. Probably the most effective remedy is to suck off the boundary layer just upstream of the point of shock impingement. The shock then effectively stands on a surface free of boundary layer as indicated in figure 4.21.

Suction can also be used to control the boundary layer in the convergent-divergent portion of the diffuser downstream of the shock system. The amounts of flow that must be taken through the suction slot and porous wall are very small compared to the throughflow of the diffuser—of the order of a few percent.

4.2.2.6 Off-Design Behavior   A fixed-geometry inlet has at least three important operating variables: flight Mach number, angle of attack, and mass flow (or the back pressure imposed by the engine). Leaving the effects of angle of attack aside for the moment, we consider the effects of varying Mach number and mass flow on the two simple types of diffuser, internal compression, and external compression.

As the flight Mach number is increased from zero for a simple diffuser of the type shown in figure 4.15, a normal shock forms at $M_0 = 1$ and gradually moves toward the lip as M_0 approaches M_{0c}. At M_{0c} it pops through the convergent section, provided the downstream condition is such that the inlet can pass the full mass flow. If we imagine this downstream condition to be imposed by a variable area nozzle as in figure 4.15d and it is choked, the requirement on the nozzle area is that

$$A_n(\rho u)_n^* = A_c(\rho u)_0^* A^*/A(M_0). \tag{4.18}$$

Now if the shock moves downstream in the divergent passage, it occurs at a higher M and so produces a larger stagnation pressure loss, reducing $(\rho u)_n^*/(\rho u)_0^*$ and requiring larger A_n. By adjusting A_n to the above value the shock can be put near the throat, and the diffuser then has the pressure recovery given in figure 4.16. Increasing M_0 beyond the starting value M_{0c} while holding the shock at the throat results in somewhat poorer pressure recovery than that in figure 4.16 because the contraction ratio A_0/A_t is smaller than it should be for best π_d. For this off-design condition π_d is the stagnation pressure ratio for a normal shock at the throat Mach number M_t found from

$$A(M_t)/A^* = [A(M_0)/A^*]A_t/A_c. \tag{4.19}$$

As an example, consider an inlet with $M_{0c} = 3$. From figure 4.16, A_c/A_t should be 1.38. Using this relation and figures 4.1 and 4.2 we find π_d to be as shown in figure 4.22. As M_0 increases from 1, π_d corresponds to a normal shock until the inlet "starts" at M_{0c}, when it is given by a normal shock at M_t. If M_0 is reduced while the shock is kept at the throat by adjusting A_n, the shock weakens until $M_t = 1$. Below this value of M_0 (1.72 for this case) the throat will not pass the full capture mass flow, so a spill shock forms ahead of the inlet.

With the shock at the throat, the inlet is unstable in the sense that any perturbation moving the shock slightly upstream will cause it to "pop" forward leading to unstart. It is therefore necessary to operate with the shock a bit downstream of the throat to ensure stability, and this lowers π_d somewhat.

Mass flow variations at fixed M_0, as might be caused (schematically) by variation of A_n, also cause changes in the shock system and variations in π_d. Consider first operation at $M_0 = M_{0c}$, so that the inlet can swallow the shock if A_n is increased to a large enough value. If A_n is reduced below the value that places the shock at the throat, the shock will pop in front of the inlet, and π_d will be reduced to that for a normal shock at M_0. It will retain this

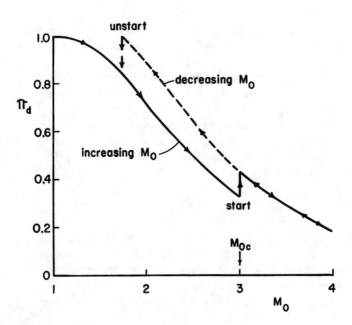

4.22 Off-design behavior (with Mach number) of simple internal compression diffuser of figure 4.15, showing hysteresis connected with starting.

value as A_n is further reduced, with a corresponding reduction in mass flow. The excess mass flow is spilled over the lip behind the normal shock. This mode of operation is termed "subcritical."

If A_n is increased from the critical value, the mass flow cannot increase because the flow is supersonic up to the throat. To accommodate the increased A_n the shock will move downstream into the divergent portion of the diffuser, where the flow is now supersonic, until it stands at a high enough M to produce the reduction in $(\rho u)^*$ required to match the larger A_n. The pressure recovery will then be given by

$$\pi_d = (\pi_d)_c A_{nc}/A_n \tag{4.20}$$

where the subscript c denotes the condition with the shock at the throat. This mode of operation is termed "supercritical." Figure 4.23 shows this behavior at $M_0 = 3$ for an inlet with $M_{0c} = 3$. At higher M_0 the behavior is similar because the inlet can start as A_n is increased through the value $A_{nc}(M_0)$ that puts the shock at the throat.

For $M_0 < M_{0c}$, on the other hand, there is an hysteresis because the inlet will not start if A_n is increased continuously from a small value. The bow shock will remain in front of the inlet, and when A_n reaches A_{nc} a weak shock will form at the (now sonic) throat and increase in strength as it

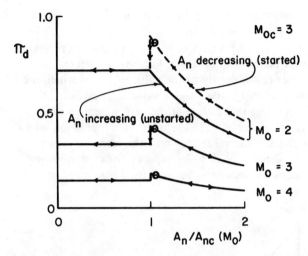

4.23   Off-design behavior with changing mass flow for the simple internal compression inlet of figure 4.15, showing double-valued (started and unstarted) operation for $M_0 < M_{0,c}$.

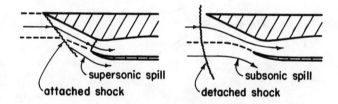

supersonic spill
attached shock

subsonic spill
detached shock

4.24   External compression diffuser at flight Mach number below design, with attached shock and supersonic spill (left) and with detached shock and subsonic spill (right).

moves downstream with further increasing A_n. If the inlet has been started as in figure 4.22, by going to higher M_0 for example, it operates on the upper curve; if A_n is reduced below A_{nc}, it unstarts and reverts to the lower curve.

The most desirable operating point for a diffuser of this type is just above critical, as marked by the circles on figure 4.23.

External compression inlets do not suffer the complications of starting and unstarting; thus their behavior is somewhat simpler. Consider, for example, the behavior of the diffuser of figure 4.17 as M_0 is reduced from the design value of 4. Reference to figure 4.4 shows that with fixed δ the shock angle θ increases as M_0 decreases, so the shock stands off the lip. Because the flow is deflected by the wedge behind this shock, the captured streamtube becomes smaller, as indicated at the left in figure 4.24. The excess mass flow is spilled by *supersonic turning* over the lip. With the proper back pressure, the second shock can be positioned at the lip.

To estimate the pressure recovery, the shock angle is found from figure 4.4 and M_1 from figure 4.5. The stagnation pressure recovery for each shock is taken from figure 4.2, and π_d is the product of the two stagnation pressure ratios. The result is shown at the bottom of figure 4.25. As M_0 decreases, the pressure recovery increases continuously to the value of M_0 at which the oblique shock detaches from the wedge. Below this value π_d is essentially that for a normal shock inlet. Note that although the detachment causes a discontinuity in π_d, there is no hysteresis as in the internal compression inlet.

A reduction in nozzle area from the value that puts the second shock at the lip, called A_{nc}, causes the second shock to be expelled forward, allowing subsonic spill. Ideally this does not reduce π_d because the Mach number is

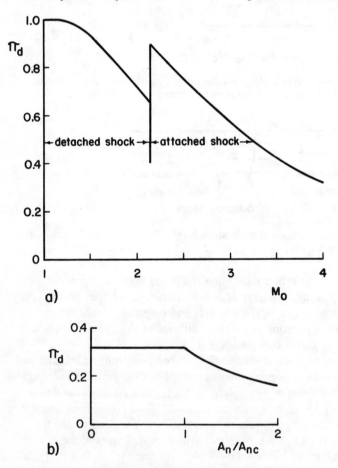

4.25 Off-design performance of external compression inlet, (a) Variation of π_d with M_0. (b) Variation of π_d with nozzle area.

uniform in the region behind the first shock. Increasing A_n from A_{nc} does not increase the mass flow, but it decreases π_d as for the external compression inlets.

4.2.2.7 Variable-Geometry Diffusers   The effects of diffuser pressure recovery and mass flow on propulsion system performance are so strong that considerable complexity and cost are justifiable in modern high-performance aircraft to bring these parameters as close to the optimum as possible. Because the requirements vary in detail with the mission of the aircraft, this discussion will consist only of some examples.

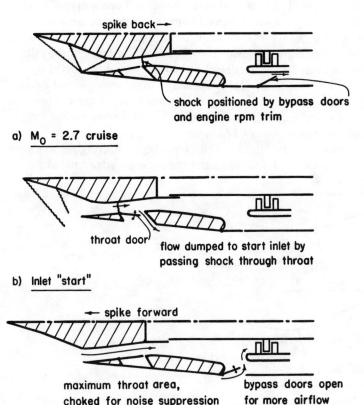

4.26   Variable-geometry mixed-compression inlet, (a) In supersonic cruise configuration. (b) During inlet "start", with throat doors open to allow shock to pass throat. (c) At takeoff, with bypass doors open for more air flow, and throat choked to suppress compressor noise.

The type of inlet proposed for the Boeing 2707–300 (SST) is shown schematically in figure 4.26. It had a double-angle conical spike, with boundary
layer suction at shock impingement points. The spike translated fore and aft
to vary the effective throat area. There was more internal contraction than
would permit starting of a fixed geometry inlet. To swallow the shock,
throat doors were opened, bypassing some flow overboard. Under some conditions a part of the flow was bypassed around the engine (called secondary
air) directly to the nozzle. Three configurations for the inlet are shown
schematically in figure 4.26. At the top is the $M_0 = 2.7$ cruise configuration,
with the last shock positioned just downstream of the throat by engine speed
and bypass door adjustment. At the bottom is the takeoff configuration.
Here the spike is fully forward, to give maximum throat area because the
engine is operating at high flow Mach number. An important feature is that
the throat is choked, so that compressor noise propagation through the inlet
is greatly reduced. At some M_0 exceeding 1.6 the inlet would be started by
retracting the spike with the throat doors opened to allow enough spill upstream of the throat so the shock could pass through, as shown at the center.

A second type of variable-geometry inlet, using external compression, is
shown configured for three values of M_0 in figure 4.27. The lower sketch
depicts the design condition at $M_0 \approx 2$ with three shocks converging on the
lip. At lower M_0 the spike is slid forward and the second wedge rotated to

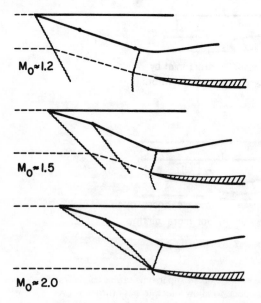

4.27   Schematic of variable-geometry external compression inlet with sliding wedge, and
variable angle on second half of wedge.

a lower angle; at the slightly supersonic condition $M_0 \approx 1.2$ it is straight, at the low angle of the front wedge. This prevents shock detachment at this low M_0.

4.3 Exhaust Nozzles

Like the inlet, the exhaust nozzle increases in complexity as the maximum flight Mach number of the aircraft engine increases, with resultant increases in nozzle pressure ratio and nozzle exit Mach numbers. The nozzle pressure ratio for a simple turbojet varies, from a value between 2 and 3 at takeoff to as large as 40 at $M_0 = 3$. To achieve perfect expansion at all flight conditions, the nozzle would have to be convergent-divergent and capable of a wide range of variations of the ratio of exit area to throat area. In addition, the throat area required for best engine performance changes somewhat with flight Mach number. It increases by a factor of 1.4 or so when the engine changes to afterburning operation.

These trends are easily seen from the ideal cycle analysis of the simple turbojet. The ratio of exit to throat area of the nozzle is determined by the nozzle exit Mach number M_7 given by (see section 2.3)

$$M_7^2 = [2/(\gamma - 1)](\theta_0 \tau_c \{1 - [(\theta_0/\theta_t)(\tau_c - 1)]\} - 1). \tag{4.21}$$

The area ratio is then $A(M_7)/A^*$. If the nozzle throat is choked, as is almost always the case for turbojets, the ratio of nozzle throat area to compressor inlet area is $A_n/A_2 = [(\rho u)_2^*/(\rho u)_6^*]A^*/A(M_2)$. Noting that $(\rho u)^* \propto p_t/\sqrt{T_t}$ and taking the pressure and temperature ratios from section 2.3, gives

$$\frac{A_n}{A_2} = \left(\frac{\theta_t}{\theta_0 \tau_c}\right)^{1/2} \left(\tau_c \left\{1 - \frac{\theta_0}{\theta_t}(\tau_c - 1)\right\}\right)^{-(\gamma+1)/2(\gamma-1)} \frac{A^*}{A(M_2)} \quad \text{(nonafterburning)},$$

$$\frac{A_n}{A_2} = \left(\frac{\theta_a}{\theta_0}\right)^{1/2} \left(\tau_c \left\{1 - \frac{\theta_0}{\theta_t}(\tau_c - 1)\right\}\right)^{-\gamma/(\gamma-1)} \frac{A^*}{A(M_2)} \quad \text{(afterburning)}.$$

$$\tag{4.22}$$

These results are shown in figure 4.28 for a turbojet with $\pi_c = 12$, $\theta_t = 7.5$. From the right figure, the ratio of nozzle throat area A_n to compressor inlet area A_2 is nearly constant up to $M_0 = 2$, then rises sharply. The assumption that M_2 is constant with varying M_0 is equivalent to assuming the corrected weight flow $W_2\sqrt{\theta_2}/\delta_2$ of the engine is constant, which in turn implies $N/\sqrt{\theta_2}$ is constant (see chapter 5). Ordinarily engines are operated at more nearly fixed N, so as θ_2 increases with increasing M_0, $N/\sqrt{\theta_2}$ decreases and so does M_2, tending to decrease the required A_n/A_2 somewhat.

The ideal nozzle area ratio A_7/A_n changes markedly with M_0, becoming as large as 7 at $M_0 = 3$. As indicated in figure 3.3 the loss associated with

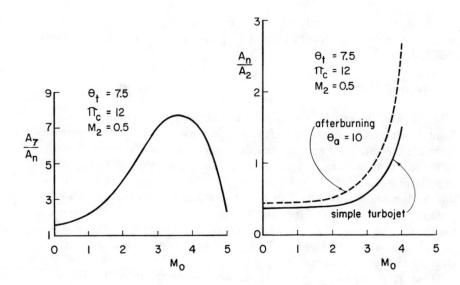

4.28   Nozzle throat area and area ratio trends with flight Mach number for turbojets.

use of a simple convergent nozzle ($A_7/A_n = 1$) is fairly small for M_0 up to 1, but is 14 percent at $M_0 = 1.5$ and more than 50 percent at $M_0 = 3$, so it is essential to use a convergent-divergent nozzle on supersonic aircraft.

4.3.1   Fixed Nozzles

A simple convergent-divergent internal expansion nozzle is shown schematically in figure 4.29. Having an area ratio of 4, this nozzle might operate ideally on a turbojet at $M_0 \approx 2$, as in sketch c. At higher M_0, it would have too low an area ratio, so that the flow would expand outside the nozzle through an expansion fan centered on the nozzle lip, as shown in sketch d. For values of M_0 less than 2 the nozzle is overexpanded, that is, the pressure at the nozzle exit plane is less than ambient pressure if the nozzle flows full. If the pressure at the exit plane is greater than about half the ambient pressure, compression through an oblique shock adjusts the pressure, as indicated in sketch b. At yet lower M_0 where the pressure ratio required of the exit shocks in configuration b would exceed about 2, the boundary layers separate from the nozzle walls as a result of the sharp pressure rise across the shock, so that the nozzle no longer flows full, as shown in sketch a. The pressure at the wall in the portion of the nozzle downstream of the separation point is nearly ambient, so the thrust of the nozzle in this flow configuration is nearly that of a nozzle of area ratio A_s/A_n, where A_s is the area at the separation point.

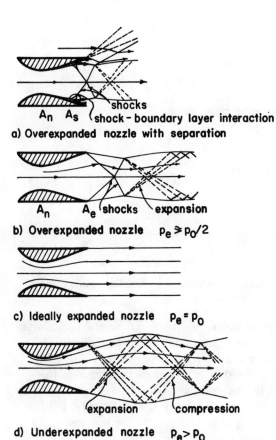

a) Overexpanded nozzle with separation

b) Overexpanded nozzle $p_e \gtrsim p_0/2$

c) Ideally expanded nozzle $p_e = p_0$

d) Underexpanded nozzle $p_e > p_0$

4.29 Design point (case c) and off-design flow in a convergent-divergent nozzle.

The thrust of the nozzle can be estimated very well for any of these operating conditions from (1.12), provided A_e is interpreted as A_s and p_e as the pressure at A_s in case a and provided the external flow is known, so that the effective ambient pressure can be determined. However, the external flow interacts with the internal flow. If this interaction influences either the internal flow or the external flow upstream of the nozzle exit plane as in case a, then neither the drag nor the thrust can be determined without understanding the interaction.

For the overexpanded nozzle (case a) the pressure at separation A_s is controlled by the static pressure rise in the shock–boundary layer interaction. For turbulent boundary layers this pressure ratio is about 2, so that as a rule of thumb $p_s = p_e \approx p_0/2$ for this case, where p_0 is the effective ambient pressure.

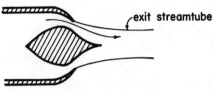

a) Overexpanded nozzle (pressure ratio below design)

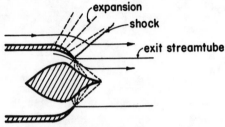

b) Ideally expanded nozzle

4.30   Design point (case b) and overexpanded operation of an external expansion "plug" nozzle.

Just as the off-design problems of supersonic inlets can be eased somewhat by utilizing external compression, the nozzle problem can be reduced by incorporating external expansion in the design. The plug nozzle shown in figure 4.30 is the idealized example of this. It is the exact analogue of the isentropic spike inlet of figure 4.18. Because the flow is expanding rather than compressing, less care need be given to the form of the centerbody; but for the same pressure ratio the turning must be the same in the two cases, and where the inlet had a serious lip drag problem, the nozzle has a serious base drag problem due to the expansion around the nozzle shroud.

The improvement in off-design performance results from the flow remaining attached to the spike at pressure ratios below design, while the streamtube leaving the nozzle contracts to satisfy the requirements for lower expansion ratio. This mode of operation is shown in figure 4.30a.

Cooling the plug is a serious problem with the plug nozzle when used on afterburning engines. It has not been much used in the simple form of figure 4.30, but external expansion has been incorporated in the integration of the nozzle with the afterbody of the aircraft. An example of this is the McDonnell-Douglas F-4, in which the nozzle exhaust expands against the aft underbody of the fuselage.

4.3.2   Variable-Geometry Nozzles
Most supersonic aircraft use afterburning for supersonic flight. Because they are also usually required to operate without afterburning for subsonic cruise

and landing, at least a two-position variation of the nozzle throat area is required. The variation can be achieved by closing a pair of eyelid-shaped segments over the end of the larger nozzle or by a series of overlapping leaves.

Ejector nozzles used on many high performance aircraft are designed so that a secondary airflow provides an aerodynamically variable expansion ratio. Two implementations of this idea are shown schematically in figure 4.31. In either case the secondary air, which may have been bled from the inlet to prevent boundary layer separation or allowed to bypass the engine to prevent the inlet shock from "popping," flows over the outside of the afterburner, cooling it, and then passes through the outer part of the nozzle. Changing the amount of secondary air varies the nozzle area available to the engine airflow, thus varying the effective expansion ratio of the nozzle. To prevent separation in the divergent portion of the nozzle, it may be contracted for low M_0 operation, as indicated in figure 4.31.

At takeoff and low M_0 conditions the nozzle may require more air than is available from the inlet. Blow-in doors just ahead of the nozzle throat may then be added to provide a tertiary air flow which helps to fill the nozzle as shown in figure 4.31b. These doors operate in response to the pressure difference across them, so that they essentially prevent the pressure in the nozzle from falling below the ambient static pressure. This reduces the drag caused by overexpansion. At higher Mach numbers, when the ram pressure rise raises

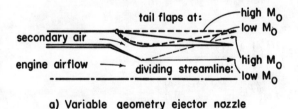

a) Variable geometry ejector nozzle

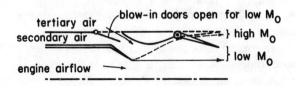

b) Ejector nozzle with blow-in doors for tertiary air

4.31 Schematics of two types of "ejector" nozzles in which secondary airflow is used to vary expansion ratio of nozzle.

the pressure of the secondary air above the ambient, the blow-in doors close, and the nozzle becomes a two-stream ejector like figure 4.31a.

As a first approximation, the ejector nozzle can be analyzed by assuming that the secondary and primary air flows are isentropic and that their static pressures are equal at each axial station of the nozzle. Thus, if we denote by subscript 1 the primary flow and by subscript 2 the secondary flow,

$$p_1 = \frac{p_{t1}}{(1 + \frac{1}{2}(\gamma - 1)M_1^2)^{\gamma/(\gamma - 1)}} = p_2 = \frac{p_{t2}}{(1 + \frac{1}{2}(\gamma - 1)M_2^2)^{\gamma/(\gamma - 1)}}$$

or

$$\frac{1 + \frac{1}{2}(\gamma - 1)M_2^2}{1 + \frac{1}{2}(\gamma - 1)M_1^2} = \left(\frac{p_{t2}}{p_{t1}}\right)^{(\gamma - 1)/\gamma} \tag{4.23}$$

The area of the primary flow streamtube is given as a function of M_1 by

$$\frac{A_1}{A_1^*} = \frac{1}{M_1}\left(\frac{1 + \frac{1}{2}(\gamma - 1)M_1^2}{\frac{1}{2}(\gamma + 1)}\right)^{(\gamma + 1)/2(\gamma - 1)} \tag{4.24}$$

and A_1^* may be interpreted as the primary nozzle throat area. The flow area of the secondary streamtube is determined from its mass flow and stagnation conditions relative to the primary streamtube,

$$\frac{A_2}{A_1^*} = \left(\frac{A_1}{A_1^*}\right)\left(\frac{\dot{m}_2}{\dot{m}_1}\right)\left(\frac{p_{t1}}{p_{t2}}\right)\left(\frac{T_{t2}}{T_{t1}}\right)^{1/2}\left(\frac{M_1}{M_2}\right)\left(\frac{1 + \frac{1}{2}(\gamma - 1)M_2^2}{1 + \frac{1}{2}(\gamma - 1)M_1^2}\right)^{(\gamma + 1)/2(\gamma - 1)},$$

and using (4.23) gives

$$\frac{A_2}{A_1^*} = \left(\frac{\dot{m}_2}{\dot{m}_1}\right)\left(\frac{p_{t1}}{p_{t2}}\right)^{(\gamma - 1)/2\gamma}\left(\frac{T_{t2}}{T_{t1}}\right)^{1/2}\left(\frac{1 + \frac{1}{2}(\gamma - 1)M_1^2}{\frac{1}{2}(\gamma + 1)}\right)^{(\gamma + 1)/2(\gamma - 1)}$$
$$\cdot \left[\frac{2}{\gamma - 1}\left((1 + \frac{1}{2}(\gamma - 1)M_1^2)\left(\frac{p_{t2}}{p_{t1}}\right)^{(\gamma - 1)/\gamma} - 1\right)\right]^{-1/2}.$$

The total flow area is then given as a function of M_1 by $A/A_1^* = (A_1/A_1^*) + (A_2/A_1^*)$. Figure 4.32 shows the relative variations of A_1/A_1^* and A_2/A_1^* with M_1 for $p_{t1}/p_{t2} = 3$, a value typical of high performance turbojets. Note that the secondary air stream occupies less area relative to the primary air as M_1 increases. This results in an expansion ratio for the primary jet that is larger than the area ratio of the nozzle. Indeed A_2/A_1^* goes to infinity at the value of M_1 where the stagnation pressure of the secondary air equals the static pressure of the primary air. Of course, the ejector would not work below this value of M_1. The changing ratio of A_2 to A_1 is illustrated by the points in figure 4.32, which are plotted for $(\dot{m}_1/\dot{m}_2)(T_{t1}/T_{t2})^{1/2} = 1$.

For a given variation of total flow area along the nozzle, the area available to the primary jet can obviously be varied by changing $\dot{m}_2/\dot{m}_1$.

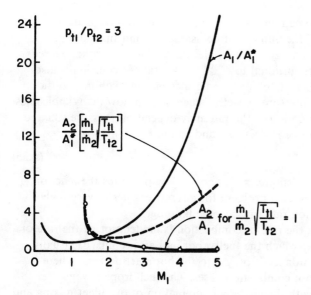

4.32   Ratios of flow areas for primary and secondary airflows to primary throat area for an ejector nozzle with engine pressure ratio = 3.

4.4   Combustors and Afterburners

The combustor must permit air at compressor outlet conditions to react with the fuel, usually a liquid hydrocarbon similar to kerosene, where the overall mixture of fuel to air varies over a wide range with changing engine speed and the compressor outlet air varies in pressure and temperature with altitude and flight Mach number as well as with engine speed. High air mass flow densities in modern compressors dictate similarly high flow densities in the combustor, and these in turn imply volumetric rates of heat release much larger than those realized in most combustion systems.

A thorough discussion of the details of the combustion process is beyond the scope of this book, but some comments are essential for understanding the behavior of combustors. The actual reaction of fuel with air takes place in the *gas phase*. The fuel vapor must be mixed *on a molecular scale* with the air at a temperature that leads to a high chemical reaction rate. The reaction rate depends in a complex way on the temperature, but for many situations it can be approximated by an Arrhenius form such as

$$\text{reaction rate} \propto f(T)e^{-A/RT}$$

where A is called the activation energy and is typically of the order of 60 kcal/mole. Physically, this expression stems from two facts: first, two molecules must collide with some minimum energy in order to react and, second, in a gas at temperature T the number of collisions per unit time and per unit volume in which the energy of one molecule relative to another exceeds a particular value A is proportional to $e^{-A/RT}$. This rate of collision is also proportional to the number of colliding molecules per unit volume, so the reaction rate depends on pressure as well. Depending on how many molecules are involved in a reactive collision, the rate may depend on pressure squared (two molecules), cubed (three molecules), and so on, so we may write

$$\text{reaction rate} \propto p^n f(T)e^{-A/RT}. \tag{4.25}$$

For hydrocarbon-air combustion $n \approx 2$, and for low pressures the reaction rate becomes slow. This poses a problem for aircraft engines at very high altitudes, where the reaction rate can become limiting.

Under most conditions the rate of combustion is limited not by the rate of reaction but by the rate at which the fuel vapor and air are mixed. Usually, the fuel is injected as an atomized spray into the hot reaction zone where it mixes with air and with hot combustion gases. The fuel droplets vaporize; the vapor is then mixed with the air by a combination of turbulent mixing and molecular diffusion. If the temperature and pressure in the reaction zone are sufficiently high, the reaction rate will not be limiting and the fuel vapor will react as soon as it comes in contact with sufficient oxygen.

Fundamental studies of this complex process necessarily idealize it. They can be divided into studies of combustion in premixed gases and studies of diffusion flames. Some ideas drawn from these studies will be summarized here.

4.4.1 Combustion in Premixed Gases

Suppose that a uniform mixture of fuel vapor and air has been formed, at some initial temperature T_0 and pressure p_0 and that ignition is attempted, for example by an intense electric spark; then we may ask two questions: (1) Does the mixture ignite and continue to burn? (2) At what rate does the flame propagate?

The answer to the first question is that at usual pressures and ambient temperatures, hydrocarbon-air mixtures will react only over a rather narrow range of fuel/air ratios—from about 0.9 to 1.2 of stoichiometric at atmospheric pressure and not at all below about 0.2 atm at standard temperature. The equivalence ratio is defined as the fuel/air mass ratio, divided by the fuel/air ratio required for complete combustion. For hydrogen, the reaction

$$2H_2 + O_2 \rightarrow 2H_2O$$

requires 4 kg of H_2 per 32 kg of O_2 or 4 kg of H_2 per $32/0.23 = 139$ kg of air, so the stoichiometric fuel air ratio for H_2 is $4/139 = 0.0288$. For octane

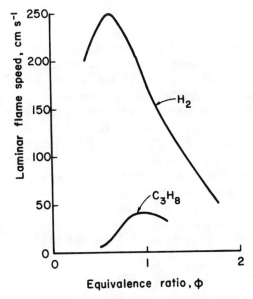

4.33   Laminar flame speed as a function of equivalence ratio for propane and hydrogen (from reference 4.5).

(C_8H_{18}) it is 0.0667. In chapter 2 we found that the fuel/air ratio required to give the desired levels of turbine inlet temperature is below about 0.03, corresponding to $\phi \approx 0.5$, so that it is not possible simply to thoroughly premix the fuel and air in a combustor and then react the mixture. Rather the fuel must be mixed with part of the air and then burned, and the combustion products must be diluted with the remaining air.

Hydrogen has much wider flammability limits than hydrocarbons, roughly $0.25 < \phi < 6$ at 1 atm and standard temperature, and some special, rather expensive fuels have flammability limits intermediate between hydrogen and the heavy hydrocarbons. They have been used at times for testing and to extend the altitude limits of engines for special applications.

If the mixture does ignite, we may then ask at what rate the flame propagates into the gas mixture. This question has been the subject of a great deal of theoretical and experimental study, and the speed can be predicted with reasonable accuracy if the flow is *laminar*. The rates of propagation for hydrocarbon-air mixtures are of the order of 30 cm s^{-1} near $\phi = 1$, dropping off very rapidly for $\phi < 0.5$ as shown in figure 4.33 for C_3H_8. If the gas mixture is *turbulent*, the rate can be as much as ten times larger, or 3 m s^{-1}, evidently because the turbulence mixes hot combustion gases with the colder fuel-air mixture. Both speeds are much higher for H_2; the laminar value is shown in figure 4.33.

4.4.2 Diffusion Flames

The most common example of a diffusion flame is the candle. Wax vaporizes from the wick and diffuses outward to meet air diffusing inward; the result is a flame front stationary in space. Even this simple example is complicated greatly by natural convection currents set up by the gas heating. A fuel droplet shot into an oxidizing, hot gas behaves in a similar fashion to the candle flame. The vaporized fuel diffuses outward to meet the inward diffusing oxygen, but the process is confused by motion of the droplet through the gas. Studies on spray combustion have shown that droplets decrease in diameter roughly according to $D^2 = D_0^2 - kt$ where D is the drop diameter, D_0 is the original diameter, and k is of the order of 0.01 cm^2/sec (see reference 4.6). If the droplets are to reside in the combustion chamber only 10^{-3} sec, say, their diameter should be less than $D_0 \approx (kt)^{1/2} \approx 0.3 \times 10^{-2}$ cm to be completely vaporized.

In some modern combustors the fuel is vaporized prior to injection to overcome the limitations of spray injection.

4.4.3 Primary Combustors

A typical combustor is shown schematically in figure 4.34. The upper half of the figure shows the way a recirculation zone is created by a swirl in the primary air and a high-velocity jet toward the burner axis part way downstream, and the way the combustion products are then diluted with air before entering the turbine nozzles. The lower half shows one method of cooling the burner "can" by a film of air introduced through slits. The slits, of course, are arranged to cool the entire circumference of the can, and the injection holes are arranged symmetrically on top and bottom or around the circumference if the can is circular. Two configurations of burners have been used. In one a series of cylindrical cans, of cross-section as in figure 4.34, are

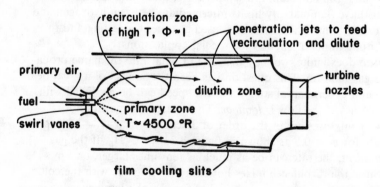

4.34 Typical gas turbine combustor showing primary zone of near-stoichiometric combustion followed by dilution with excess air to reach T_{t4}.

arranged around the circumference of the engine. In the other a single annular burner may be used. Figure 4.34 would then be a cross-section in a plane through the engine's axis; there would also be several sets of nozzles, but not necessarily the swirl vanes.

The gas temperature in the primary zone is much higher than the materials (such as Hastelloy X) of which the burner is constructed can tolerate, so that the cooling must be very effective. Further, the temperature profile in the radial direction at burner exit must be well controlled under all operating conditions to minimize the stress problems in the turbine. This temperature profile is sometimes deliberately made nonuniform, with a lower temperature at the blade roots where the stress is ˙argest and at the tip, because the outer wall is hard to cool.

Two principal performance parameters for the burner are the efficiency η_b of conversion of the fuel's chemical energy to heat and the stagnation pressure drop π_b. The pressure drop depends on viscous processes such as the flow through the orifices and cooling slits of the burner and on the density changes produced by combustion. It can therefore be correlated in terms of the flow Mach number at entrance to the burner and the stagnation temperature ratio (plus the Reynolds number for some situations). A useful rule of thumb is that the stagnation pressure drop is about one or two times the flow dynamic pressure based on the flow area of the combustor. A little ciphering will show that this is equivalent to

$$\pi_b \approx 1 - \varepsilon(\gamma/2)M_b^2 \tag{4.26}$$

where $1 < \varepsilon < 2$ and M_b is the Mach number based on burner flow area. For example, if $M_b \approx 0.2$ and $\varepsilon = 2$, then $\pi_b \approx 0.94$, a typical value.

Unfortunately, the number of physical processes that influence the combustion process is so large that despite many attempts, no system for correlating burner efficiency in terms of dimensionless parameters has been evolved. This is easily understood when one realizes that atomization depends on fuel surface tension, injector pressure drop, and shape, that the vaporization process depends on the fuel's vapor pressure and heat of vaporization, that the reaction itself depends on the activation energy and chemical composition, and so on. Some general observations can be made. Because reaction rate depends on temperature and pressure, efficiency tends to decrease with decreasing pressure and inlet air temperature. It also tends to decrease with increasing combustor inlet Mach number M_b because this decreases the residence time of the fuel-air mixture in the burner. Finally, because of the sensitivity indicated by figure 4.34, η_b tends to decrease as the fuel/air ratio is varied from the nominal design value.

Some qualitative trends can be identified by considering a family of combustors identical except for different pressures and mass flows if we adopt the view that the efficiency depends on the residence time in the combustor. This residence time is $\tau_{res} \approx A_b \rho_{t3} L / \dot{m}$, where L is the burner length and A_b its flow area, and ρ_{t3} is the compressor outlet density. Rewriting this somewhat gives

$$L \propto [(\dot{m}/A_2)/(A_b/A_2)][\tau_{res}/(\pi_c)^{1/\gamma}]. \tag{4.27}$$

If we keep the ratio of burner area to compressor frontal area constant and the compressor mass flow per unit frontal area constant, then $L \propto \pi_c^{-1/\gamma}$. If we hold π_c constant but change the *size* of the engine, L remains constant, so the ratio of combustor length to engine diameter decreases as the engine size increases. For both of these reasons, combustor size decreases relative to engine size in large, high pressure ratio engines compared to small, low-pressure engines. This can be seen very clearly by comparing figures 1.15 and 1.19.

4.4.4 Afterburners

The higher temperatures and near stoichiometric fuel/air ratios of after-burners enable them to operate with a simpler configuration than the primary combustor. On the other hand, the relatively high flow Mach number makes pressure losses somewhat more critical in the afterburner. The Mach number can be estimated from figure 4.28, which gives the ratio of nozzle throat area to compressor inlet area for a simple turbojet. If we assume the afterburner flow area equals the compressor flow area, $A_6 = A_2$ (see figure 1.4), then $A(M_6)/A^* = A_2/A_n$, and with the values of A_n/A_2 from figure 4.28 we can find M_6 from figure 4.1 for either the afterburning or nonafterburning case. We can take the nonafterburning value as the Mach number before combustion in the afterburner and the afterburning value as that after combustion. These values are shown in figure 4.35. Above $M_0 = 3.2$ this estimate shows the afterburner to be "thermally choked," and either A_6 would have to be increased or θ_a reduced.

A typical afterburner configuration is shown in figure 4.36. The fuel is sprayed into the turbine exit annulus where it vaporizes and mixes. A flame-holder is provided to stabilize the flame front, which stands at an angle to the flow determined by the flow velocity and the propagation velocity of the turbulent flame front, just as the angle of an oblique shock is determined.

The afterburner flameholder stabilizes the flame front by producing a region of recirculating flow in which there is a large residence time. According to the model developed in reference 4.7 and sketched in figure 4.36, the recirculation zone is surrounded by a mixing zone, and the residence time of the fluid in this mixing zone determines whether a stable flame will form or whether the flame will "blow off." Because of turbulent mixing, burned and unburned material enter the mixing zone from the unburned flow and

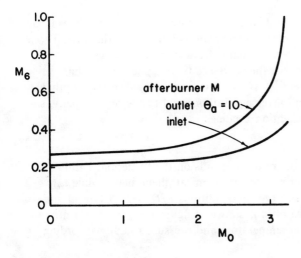

4.35   Afterburner inlet and outlet Mach numbers for the turbojet engine of figure 4.28 as functions of flight Mach number.

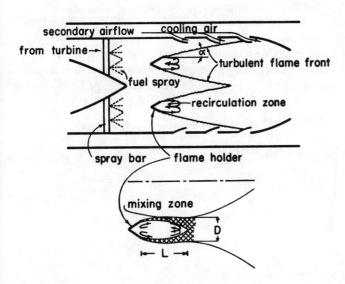

4.36   Schematic of an afterburner, showing V-gutter flame holders stabilizing flame by producing a recirculation zone (lower sketch).

recirculation zone, respectively. If the temperature and residence times in the mixing zone are sufficient, a chemical reaction is initiated in the mixing zone gas, and this reacting gas enters the flame front downstream. Thus the criterion for stabilization is that the residence time t_{res} be greater than some chemical reaction time t_{reac}. Now t_{res} is proportional to the length L of the recirculation zone and inversely proportional to the flow velocity, so the criterion for flame stabilization becomes

$$L/u > \text{const}(t_{reac}) \tag{4.28}$$

This relation is very useful because L depends on the flameholder dimension D and Reynolds number but not on the chemical phenomena, while t_{reac} depends on the chemistry. For typical conditions $L/D \approx 3$. A typical variation of t_{reac} with equivalence ratio is shown in figure 4.37, for $p = 1$ atm and $T = 340°$K. t_{reac} depends on temperature and pressure roughly according to

$$t_{reac} \propto T^{1.4}/p$$

for $250 < T < 400°$K and $0.1 < p < 1$ atm. For hydrogen-air mixtures, the minimum of t_{reac} is about one-tenth that for gasoline-air shown in figure 4.37.

The length of the afterburner is determined by the flame spreading angle (see figure 4.36), which is almost independent of cold gas velocity and near 3 degrees for typical conditions.

From these facts we can assemble a qualitative understanding of the compromises required in an afterburner design. From the engine cycle T, p, and M at the afterburner inlet are known. From (4.28) these determine a minimum

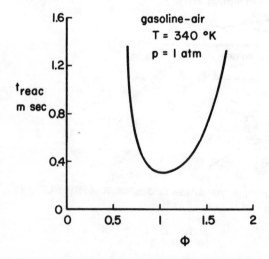

4.37   Chemical reaction time as a function of equivalence ratio ϕ for gasoline-air.

flameholder dimension D once the equivalence ratio ϕ is set. The number of flameholders that can be put in is determined by the acceptable pressure drop, and the linear spreading between them then determines the afterburner length.

Because D is determined by the engine cycle independently of engine size, the geometry of afterburners changes markedly with engine size; their length (and weight) become larger relative to the engine as the engine size decreases. A comparison of figures 1.15 and 1.19 makes this very clear.

4.4.5 Pollutant Formation

The products of combustion from either the primary burner or the afterburner would ideally be composed of nitrogen (N_2), some unconsumed oxygen (O_2), carbon dioxide (CO_2), and water vapor (H_2O). Hydrocarbon fuels always contain a little sulfur, which appears as SO_2 in the exhaust. This sulfur content must be below about 0.1 percent in aircraft fuels to avoid corrosion of the turbine blades, so the SO_2 is a very minor constituent in aircraft exhausts. It can be a major problem in stationary gas turbines where lower firing temperatures are used to accommodate the sulfur corrosion problem. Actual combustors produce a number of other products; the amount and type vary with the operating condition of the engine.

At part-throttle and idle conditions the residence time in the burner is smaller than at full throttle. This can be understood from (4.27) by noting that $\pi_c^{1/\gamma}$ increases more rapidly with increasing compressor speed than does $\dot{m}/A_2$ (again, see chapter 5). The reduced residence time leads to incomplete combustion and the exhaust of carbon monoxide (CO) and unburned hydrocarbons. Small flow rates of fuel through the fuel injection nozzles tend to lead to poor atomization, further decreasing the combustion efficiency.

At full-throttle conditions, there is a tendency toward soot (smoke) formation if the primary combustion zone is fuel rich ($\phi > 1$). This problem can be solved by leaning out the primary zone, at some cost in stability and operating range of the combustor. Most modern gas turbines do not smoke visibly.

A more fundamental and intractable problem than any of these is the production of nitrogen oxides (NO_2, NO) due to the reaction of nitrogen with oxygen in the high temperature zones of the combustor. This occurs by the set of reactions

$$N_2 + O \rightleftarrows NO + N$$

$$N + O_2 \rightleftarrows NO + O$$

$$N + OH \rightleftarrows NO + H$$

where the O, H, and OH are produced by dissociation of O_2 and H_2O. The first reaction is endothermic (an energy input is required), so its rate is given by an expression like (4.25) with a very large value of A. This means it proceeds only at very high temperatures and that the change of reaction rate with temperature is very rapid. The magnitude of the rate at primary zone temperatures is much smaller than that of the reactions that limit the rate of combustion of the fuel. The process of NO formation can thus be described schematically as shown in figure 4.38, where we follow a typical sample of air-fuel mixture as it progresses through the combustor. The temperature rises rapidly, as the fuel burns to produce CO, CO_2, and H_2O in a relatively short time in the primary zone of the burner. If there is some excess oxygen, the CO is oxidized to CO_2 as the gas is held at this temperature. Formation of NO also begins but at a slower rate. If the gases were held at the peak temperature for sufficient time, the NO concentration would build up to an equilibrium value for that temperature. Actually, the gases are cooled by mixing with excess air in the secondary zone of the burner, and the reduced temperature limits NO production to a value far below that for equilibrium at the primary zone temperature.

The rate of NO formation may be represented as

$$\frac{d[NO]}{dt} = 2k[N_2][O]$$

where [ ] denotes the concentration in particles per unit volume of the indicated chemical species. According to the scheme outlined above, [O] is determined by thermal dissociation of O_2, $O_2 \rightleftarrows 2O$, so that if the equilibrium

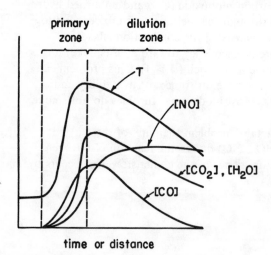

4.38   Schematic time histories of temperature and species concentrations in an air-fuel mixture passing through a combustor.

constant for this reaction is

$$K_p(T) = p_O^2/p_{O_2},$$

we find

$$\frac{1}{[N_2]} \frac{d[NO]}{dt} = \frac{2k}{RT}[p_{O_2}K_p]^{1/2}.$$

This rate of formation is shown as a function of temperature and equivalence ratio ϕ in figure 4.39 (see reference 4.8). The figure was drawn for C_nH_{2n}, a typical hydrocarbon composition, for an initial temperature of 700°K and pressure of 15 atm. The dashed line represents the temperature which would be attained in adiabatic combustion for each value of ϕ.

To interpret this figure, consider a combustor with a primary zone residence time of 1 ms (10^{-3} s) and $\phi = 1$. It would produce $[NO]/[N_2] \approx$ 0.6×10^{-3}, or about 600 parts per million of NO, and this might double or triple in the dilution zone.

It is more usual to quote the emissions in terms of grams of pollutant per kilogram of fuel burned. Because the stoichiometric-fuel/air ratio is about 0.066, the above value in these terms would be

$$\frac{\text{gm NO}}{\text{kg fuel}} \approx \left(.6 \times 10^{-3} \frac{\text{mole NO}}{\text{mole N}_2} \right)\left(\frac{30}{28}\right)\left(\frac{.77}{.066}\right)(10^3) = 7.5.$$

A typical value for current engines at full power is 20 gm NO/ kg fuel.

The proposed airport emissions standards of the Environmental Protection Agency are stated in terms of the mass of pollutant produced in a typical takeoff and landing cycle per unit of thrust. The actual unit is lb_m of pollutant per 1000 lb_f of thrust per hour of operation. The proposed values are given in the first column of the following table; values for two operational engines are given in the last two columns.

Pollutant	EPA (1979 proposed)	JT8D	JT9D
CO	4.3	19.	14.
HC	0.8	2.7	4.1
NO_x	3	8.0	8.3

The JT8D is used in nearly all short- and medium-range narrow body aircraft such as the B-727, DC-9 and B-737. It is representative of 1960 engine technology. The JT9D represents the high bypass ratio, high compression engines of the wide bodied aircraft introduced in the 1970s. For either, a factor of three reduction in NO_x will be required to meet the proposed standards.

Within the conceptual framework of the conventional combustor, the way to reduce NO_x is to reduce the residence time at high temperatures. But this must be done without decreasing the combustion efficiency appreciably or increasing CO. If the rates of combustion of CO to CO_2 and of formation of NO were both kinetically limited, a decrease of NO would definitely imply an increase of CO. But if the oxidation of CO is mixing rate limited, then increasing the mixing rate should decrease both CO and NO. Current (1976) attempts to meet the 1979 EPA standards center on burner designs with faster mixing rates, which are achieved by reducing the size of the burner. In one design (see reference 4.9) a large number of small swirl cans is introduced, effectively replacing the larger primary zone by a series of

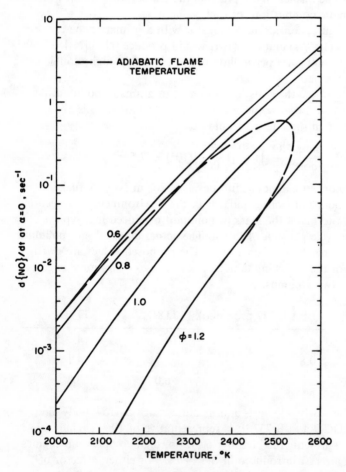

4.39 Rate of change of NO mass fraction as a function of temperature and equivalence ratio, for C_nH_{2n} initially at $700°K$ and 1 atm.

smaller ones. In another (reference 4.10) the annular burner is replaced by a double-annular burner, reducing the size by a factor of two. For either case, if the velocities are the same as in the conventional burner, the residence time is reduced by the ratio of the sizes of the primary zones.

From chapters 2 and 3, we saw that improvements in gas turbine efficiency and power output result from increased turbine inlet temperature coupled with higher compressor pressure ratios. Other things being equal, the associated increases in combustor inlet temperature result in greater NO production. A correlation of NO_x in gm NO_2 per kg of fuel versus T_{t3} shows a monotonic increase from values near 2 at 400°K to 40 at 800°K (reference 4.11) for a number of actual engines. If this trend is inevitable, it could seriously restrict the use of high pressure ratios in future engines. It is not inevitable; rather it is a result of designing for nearly constant residence time in the burners of all the engines studied, as the following argument shows.

From the kinetic argument and from figure 4.39, we deduce that the NO_x concentration in the exhaust should behave as

$$NO_x \propto p_3^{1/2} e^{-2400/T_3} t_p$$

where T_3 is in °K and t_p is the residence time in the primary zone. Now the residence time t_p is determined by the requirement for good combustion efficiency, as noted in section 4.4.3. Correlations of combustor efficiency η_b for fixed geometry combustors have indicated that η_b correlates as a function of $p_3^{1.75} \exp{(T_3/b)}/\dot{m}$ where $\dot{m}$ is mass flow through the burner and b varies from 300°K for a fuel/air ratio of 0.016 to 150°K for 0.010. Taking $t_p \propto p_3/\dot{m}$, and assuming $p_3 \propto T_3^{(\gamma-1)/\gamma}$, we find that for *constant combustion efficiency*,

$$NO_x \propto \frac{\exp{-[(2400/T_3) + (T_3/b)]}}{T_3^{\gamma/4(\gamma-1)}}, \qquad \eta_b = \text{const},$$

which is to be compared to $NO_x \propto T_3^{(\gamma-1)/2\gamma} \exp{(-2400/T_3)}$; $t_p = \text{const}$. These two variations are compared in figure 4.40 (see reference 4.12), from which we see that for combustors scaled for constant η_b, NO_x actually decreases at the higher values of T_3.

A second possibly more demanding requirement for NO_x reduction may arise from its possible impact on the ozone (O_3) layer of the stratosphere. This layer is formed by absorption of ultraviolet from the sun, and in turn shields the earth from almost all ultraviolet (wave length < 3200 Å) radiation. The series of processes supporting the O_3 layer is (reference 4.13)

$$O_3 + h\nu \rightarrow O_2 + O$$

$$O + O_2 \rightarrow O_3$$

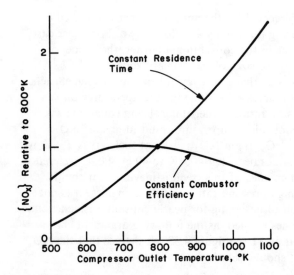

4.40 Comparison of variations of NO_x production by combustors as a function of compressor outlet temperature for constant residence time and for constant combustion efficiency.

where hv represents the ultraviolet photon while reactions such as

$$O + O \rightarrow O_2$$

$$O + O_3 \rightarrow 2O_2$$

tend to limit the O_3 concentration. Both processes are slow, and the O_3 concentration is appreciable only in the stratosphere where the increase of temperature with altitude makes the atmosphere stable to vertical disturbances in contrast to the troposphere which is thoroughly mixed vertically. In large part it is the thermal energy input due to ultraviolet absorption which in turn maintains the high temperature in the stratosphere.

The balance between these processes can be upset by introduction of NO, which acts to reduce the O_3 concentration by the reactions

$$NO + O_3 \rightarrow NO_2 + O_2$$

$$NO_2 + hv \rightarrow NO + O$$

$$NO_2 + O \rightarrow NO + O_2$$

in which no NO is destroyed, but O_3 is combined to O_2.

Estimates of the gravity of this problem vary from an upper limit of 30 percent reduction of the total O_3 cover by a fleet of 800 advanced SST aircraft to nil. The high estimate is based on NO concentrations in the exhaust of current engines, which are of the order of 20 gm NO per kg of fuel consumed.

At least two solutions to this potential problem have been suggested. One is to prohibit supersonic flight in the upper atmosphere; the other is to reduce the NO emission from the engines by a large factor. Roughly a factor of 100 would be required to cut the 30 percent potential O_3 reduction to a negligible value.

Such NO reductions are achievable, in principle, by means of a combustion process in which the temperature of the gases at no time appreciably exceeds the final turbine inlet temperature. Two possibilities have been suggested. One proposal is to take advantage of the high combustor inlet air temperature in a supersonic transport engine (perhaps as much as 1000°K) which broadens the flammability limits and allows combustion of a uniformly premixed, lean mixture. (reference 4.14) In laboratory experiments, NO concentrations as low as 1 g NO per kg fuel have been measured and calculations suggest values as low as 0.3 may be attainable. It will be a difficult task, however, to utilize this process in an engine because the combustor must be capable of operating also at low air inlet temperatures and over a range of fuel/air ratios.

The second scheme involves using solid catalysts to react the lean fuel/air mixture (reference 4.15). Again laboratory tests have been encouraging, but the problems of weight, size, and durability are formidable.

References

4.1 H. W. Liepmann and A. Roshko, *Elements of Gasdynamics*, John Wiley, 1957.

4.2 R. H. Sabersky, A. J. Acosta, and E. G. Hauptmann, *Fluid Flow, A First Course in Fluid Mechanics*, Macmillan, 1971.

4.3 W. E. Moeckel and J. F. Connors, "Charts for the Determination of Supersonic Air Flow Against Inclined Planes and Axially Symmetric Cones," NACA TN 1373, July 1947.

4.4 H. Schlichting, *Boundary Layer Theory*, McGraw-Hill, 1960.

4.5 W. Jost, *Explosion and Combustion Processes in Gases*, McGraw-Hill, 1946.

4.6 S. S. Penner, *Chemistry Problems in Jet Propulsion*, Pergamon Press, 1957.

4.7 E. E. Zukoski and F. E. Marble, "Experiments Concerning the Mechanism of Flame Blowoff from Bluff Bodies," in *Proceedings of the Gas Dynamics Symposium on Aerothermochemistry*, Northwest University Press, 1955. Also: E. E. Zukoski and F. H. Wright, "Flame Spreading from Bluff-Body Flameholders," in *Eighth Symposium on Combustion*, Buttersworth, Ltd., London, 1960.

4.8 J. B. Heywood and T. Mikus, "Parameters Controlling Nitric Oxide Emissions from Gas Turbine Combustors," AGARD Propulsion and Energetics Panel, 41st Meeting on Atmospheric Pollution by Aircraft Engines, London, England.

4.9 R. W. Niedzwiecki and R. E. Jones, "Pollution Measurements of a Swirl-Can Combustor," NASA TM Y–68160, 1972.

4.10 D. F. Schultz and D. J. Perkins, "Effects of Radial and Circumferential Inlet Velocity Profile Distortions on Performance of a Short-Length, Double-Annular, Ram-Induction Combustor," NASA TN D–6706, 1972.

4.11 F. W. Lipfert, "Correlation of Gas Turbine Emissions Data," ASME Paper No. 72–6T–60, 1972.

4.12 J. L. Kerrebrock, "The Effect of Compression Ratio on NO_x Production by Gas Turbines," *Journal of Aircraft*, Aug./Sept. 1975.

4.13 C. E. Kolb, "The Depletion of Stratospheric Ozone," *Technology Review*, Oct./Nov. 1975.

4.14 A. Ferri, "Reduction of NO Formation by Premixing," in AGARD Conference Proceedings No. 125 on *Atmospheric Pollution by Aircraft Engines*, pp. A-1–A-9, 1973.

4.15 W. S. Blazowski, and Walsh, D. E. "Catalytic Combustion: An Important Consideration for Future Applications," *Combustion Science and Technology*, 1975.

Problems

4.1 Develop the result given as (3.6) by considering flow in opposite directions along the two sides of a flat heat exchanger plate, where the flow on both sides is in constant area channels. From section 4.1.4, the rate of change of pressure along the flow direction x is $dp/dx = 2\rho u^2 C_f/d$ and the rate of change of stagnation temperature is $dT_t/dx = 4St(T_t - T_W)/d$ where T_W is the wall temperature and d is the hydraulic diameter of the flow passage.

4.2 A simple convergent-divergent internal compression diffuser operates at flight Mach numbers $0 < M_0 < 2.5$. What should its throat/capture area ratio be? What is its maximum pressure recovery at $M_0 = 2.5$? At what value of M_0 would its pressure recovery be unity?

4.3 The mixed-compression supersonic inlet shown in figure 4.19 is designed for $M_0 = 4$, where it produces $\pi_d = 0.59$ (including shock losses).
a. First calculate the losses for this point to check the above π_d.
b. Now suppose the inlet flies at $M_0 = 3$, with the same geometry. Determine the shock positions for best pressure recovery and calculate π_d for this Mach number.
c. Repeat part b for $M_0 = 2$.

4.4 Suppose that an aircraft using a simple internal compression inlet (figure 4.15) designed for $M_0 = 3$ is flying at $M_0 = 2$ with the inlet started and that the shock is optimally positioned when the inlet suddenly unstarts, popping the shock. The engine is a turbojet with $\pi_c = 12$, $\theta_t = 7.5$. By what ratio does the thrust change? To estimate this ratio assume that the engine is ideal except for the inlet pressure loss, and that the Mach number M_2 entering the engine remains fixed.

4.5 For a flight Mach number of 3, "design" a mixed compression, two-dimensional inlet such as that in figure 4.19. Draw a cross-sectional view of the inlet, showing shock positions and flow directions for the geometry that yields the best pressure recovery. Now estimate the ratio of the height of the boundary layer suction slot to the capture streamtube height, required at the point where the second shock impinges on the wedge. Assume the boundary layer is turbulent, the altitude is 10 km, and the capture height is 1 m. The momentum thickness for a turbulent boundary layer varies as $\theta^* = 0.023(Re_x)^{-1/6}x$ where x is the distance along the flat plate and Re_x is the Reynolds number based on the length x.

4.6 Starting with (4.21) compute the variation with M_0 of A_7/A_n for an ideal turbojet with $\theta_t = 7.5$, $\pi_c = 6$. Compare your results with those of figure 4.28 and explain the differences.

4.7 Carry out the preliminary design of an afterburner for a turbojet that has a corrected mass flow of 50 kg s^{-1}. At the afterburner design point, which is $h = 10$ km, $M_0 = 0.5$, the turbine exit conditions are $T_{t5} = 1000°K$, $p_{t5} = 1$ atm, and $M_5 = 0.22$. Assume that the flame holders have a drag coefficient of unity, based on the mean flow velocity at the minimum flow area and that the maximum acceptable pressure drop is one dynamic head based on the flow velocity ahead of the flame holders. To determine the flame holder dimension, take $L/D \approx 3$, $\phi = 1$, and use the data of figure 4.37.

4.8 Combustors must provide for stable operation over a range of engine speeds, from idle to full power. Estimate the ratio of the combustor inlet velocity u_3 at rated rpm to that at half of rated rpm for an engine with compressor pressure ratio $\pi_c = 8$ at rated rpm. Assume that the compression is ideal, that the compressor temperature ratio $\tau_c = 1 + c_1N^2$, where c_1 is a constant, and that $u_2 = c_2N$, where N is the rotative speed.

4.9 The afterburner on a turbojet engine is designed to operate satisfactorily at $M_0 = 0.5$ and an altitude of 20 km. Suppose now that the requirement is modified so that it must operate only up to 10 km. Could a lighter (shorter) afterburner be used? By about what factor could it be shortened?

5
Compressors

Because the compressor controls the pressure ratio of the engine, especially at low flight Mach numbers, and the mass flow as well, it has a dominant influence on the characteristics of the gas turbine engine. For this reason and because it has been one of the most difficult engine components to develop, it has received great emphasis in both research and development. Despite these efforts, which have extended over the last three decades or more, the design and development of a new compressor is still not a straightforward process. It involves sophisticated aerodynamic design, trial-and-error testing, and much intricate mechanical design; all of these are expensive and time-consuming. The improvements over the years have come, not in more straightforward design procedures, but in enormously improved performance.

Only a small part of this elaborate technology can be discussed here. Because our aim is to explain the characteristics of compressors as engine components, we will discuss only the theory and experiment essential to a qualitative understanding of the compressor's characteristics, its performance limitations, and design tradeoffs. The techniques of design to the present state of the art will be left to more advanced treatments.

Two types of compressors have been widely used in gas turbines: axial flow, in which the air flows mainly parallel to the rotational axis of the engine, and radial or centrifugal flow, in which the air is turned from the axial to the radial direction in the compressor rotor. Axial flow compressors have been used the most in aircraft engines and are the most highly developed, so they will be the main subject of discussion. Centrifugal compressors are used in small engines and for automotive applications and have received more attention recently. Their characteristics will be discussed, though not as extensively.

The feature that distinguishes compressors and turbines from the fixed aerodynamic components of the engine and aircraft is the energy exchange between rotor and fluid, which makes possible the compression and expansion required for an efficient thermal cycle. This process will be discussed first.

5.1 Energy Exchange, Rotor to Fluid

Because the process of dynamic energy exchange used in all turbomachines is not discussed in the usual courses and texts in fluid mechanics, it will be discussed from a fundamental viewpoint here. We begin with the first law of thermodynamics in the form $\delta E = \delta W - p\delta(1/\rho)$, where E is internal energy, W is mechanical work, and ρ is the density. In the inviscid limit considered here, no forces (other than the normal forces represented by p) act on the fluid, so $\delta W = 0$. Now if we consider the differentials of E and $1/\rho$ to be taken following an element of the fluid as it moves, the first law becomes

$$DE/Dt = -pD(1/\rho)/Dt \qquad (5.1)$$

where

$$D/Dt = \partial/\partial t + \mathbf{u} \cdot \text{grad}.$$

The second fundamental law is the momentum equation for the fluid. Because the force acting on the fluid per unit volume is the gradient of the pressure, Newton's second law takes the form

$$\rho(D\mathbf{u}/Dt) = -\text{grad } p. \qquad (5.2)$$

(Representing the fluid acceleration by $D\mathbf{u}/Dt$ is valid only if the velocity $\mathbf{u}$ is referred to an inertial coordinate system.)

Forming the scalar product of (5.2) with $\mathbf{u}$ gives the following expression for the rate at which the kinetic energy changes as a result of pressure forces acting on the fluid:

$$\rho D(u^2/2)/Dt = -\mathbf{u} \cdot \text{grad } p. \qquad (5.3)$$

Now (5.1) and (5.3) can be combined to give a single equation for the fluid's total energy. First introduce the enthalpy $h = E + p/\rho$ and note that $DE/Dt = (Dh/Dt) - (1/\rho)(Dp/Dt - pD(1/\rho)/Dt)$, so that (5.1) becomes

$$\rho(Dh/Dt) = Dp/Dt = (\partial p/\partial t) + \mathbf{u} \cdot \text{grad } p. \qquad (5.4)$$

Adding (5.3) and (5.4) then yields the final relation

$$\rho D[h + (u^2/2)]/Dt = \partial p/\partial t. \qquad (5.5)$$

In the approximation we have used so far, $h + u^2/2 = c_p T + u^2/2 = c_p T_t$, so (5.5) shows that *in this inviscid, non-heat-conducting limit, the stagnation temperature of the fluid, and hence its stagnation pressure, can only be changed by an unsteady compression or expansion.* No steady flow process, for which $\partial/\partial t = 0$ by definition, can affect energy addition to or removal from the fluid. Further, (5.5) shows that the energy of the fluid can be increased only by increasing the pressure. Conversely, to decrease the energy, as in a turbine, the pressure must decrease.

With these fundamental points in mind, let us consider the energy transfer between moving blades and the fluid. For the sake of this discussion, consider an axial flow compressor with such large ratio of hub diameter to tip diameter that the blades may be approximated by a linear cascade, that is, an infinite row of blades moving in a straight line, as at the top of figure 5.1. If the flow is axial and uniform ahead of the cascade, then by transforming to a coordinate system stationary in the cascade the flow seems steady, but from an angle as shown. If the blades are shaped to turn the flow toward the axis in this coordinate system (in the direction of blade motion), they form

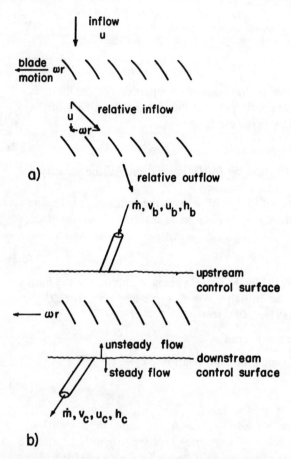

5.1 a. Cascade representation of a rotor blade row with absolute velocities (top) and relative velocities (bottom). b. The blade row and the streamtube considered in developing the Euler equation.

diverging passages which for subsonic flow result in rising pressure as the fluid passes through the cascade. The result according to (5.5) is that energy is added to the flow.

The numbers 2, 3, 4, 5, 6, 7 have been assigned to points in the engine as indicated in figures 1.4, 1.6, 2.7. To avoid confusion with these, the successive positions along the flow direction in the compressor and later in the turbine will be designated by letter subscripts a, b, c, d. Thus, if M_a is the Mach number ahead of the guide vanes, which are the first blades in the compressor, M_a is related to M_2 simply by the flow contraction caused by the rounded "spinner" in figure 1.4, for example.

5.1.1 The Euler Equation

Given the pressure distribution in the blade row, we could find the temperature and pressure ratios of the compressor by integrating (5.5). There is, however, a computationally simpler way to look at the energy transfer process, although it obscures the physical nature of the process. We draw control surfaces upstream and downstream of the cascade and suppose that the flow across these surfaces is steady, as will be true to an adequate approximation if the surfaces are far upstream and downstream of the blades. We then identify a streamtube entering the cascade through the upstream surface and emerging through the downstream one, and we apply the laws of conservation of (1) total fluid energy and (2) momentum in the direction of blade motion to this streamtube. If $\dot{m}$ is the mass flow in the streamtube, the conservation of energy requires $\dot{m}(h_c + u_c^2/2 - h_b - u_b^2/2) = P$ where P is the power delivered to the fluid in this streamtube by the blades. Conservation of momentum requires that the difference between the momentum flowing into the control volume in the upstream tube and that flowing out in the downstream tube plus the force F acting in the rotor due to streamtube sum to zero, or $\dot{m}(v_c - v_b) = F$. But if ωr is the velocity of the blades, then $P = F\omega r$, and we find

$$h_c + (u_c^2/2) - [h_b + (u_b^2/2)] = \omega r(v_c - v_b).$$

Generalizing this argument to the axisymmetric case, let the upstream tube have mean radius r_b and tangential velocity v_b, and the downstream tube r_c and v_c as in figure 5.2. The energy balance is unchanged, but the momentum balance is replaced by an angular momentum balance $\dot{m}(r_c v_c - r_b v_b) = \mathcal{T}$ where $\mathcal{T}$ is the torque due to the streamtube; now $P = \omega \mathcal{T}$, so that

$$h_c + (u_c^2/2) - [h_b + (u_b^2/2)] = \omega(r_c v_c - r_b v_b). \tag{5.6}$$

This is the famous Euler turbine equation; it will form the basis for much of the following discussion.

Putting $c_p T_t = h + (u^2/2)$, we have for a perfect gas

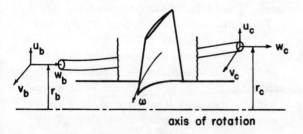

axis of rotation

5.2 The blade row, control surfaces, and streamtube for development of the Euler equation in rotor geometry.

$$c_p(T_{tc} - T_{tb}) = \omega(r_c v_c - r_b v_b). \tag{5.7}$$

Another interesting special case is the incompressible fluid. It is still true that $h = E + (p/\rho)$, but since E is not influenced by changes in pressure it is constant, and $\rho(h + u^2/2)$ becomes $p_t = p + \rho u^2/2$, so the Euler equation is

$$(p_{tc} - p_{tb})/\rho = \omega(r_c v_c - r_b v_b); \quad \rho = \text{constant}. \tag{5.8}$$

In the following sections we will assume that the flow is steady in co-ordinates moving with the blades, determine the flow patterns in these co-ordinates, and then deduce the pressure rise from (5.7) or (5.8). This classical procedure is satisfactory so long as the assumption of steady flow in rotor coordinates is justifiable.

5.1.2 Stage Temperature Ratio

There are normally three types of blade row in an axial compressor: the inlet guide vanes, the rotor blades, and the stator blades. The inlet guide vanes usually give the flow a swirl in the direction of rotor motion to reduce the flow velocity relative to the rotor blades, minimizing shock losses, and to equalize the pressure rises in rotor and stator. The rotor adds energy to the flow and in the process imparts angular momentum to it. The stator removes this angular momentum and diffuses the flow to raise the pressure. The rotor-stator combination constitutes a stage.

It is convenient to represent the changes in velocity that occur through the blading by a velocity diagram. The development of such a diagram is shown in figure 5.3 for an inlet guide vane row, plus one stage. The velocity changes across each blade row are shown at the top of the figure, in relation to the blades. The flow angle, measured from the axial direction, will be denoted β with a subscript for the axial station. Angles in the coordinate system fixed in the rotor are denoted by a prime. Velocities in the stationary coordinate system are shown by solid arrows, those in the rotor coordinate frame by dashed arrows. Thus, the inlet guide vanes turn the flow to the angle β_b and in the process raise the Mach number from M_a to M_b. The rotor blades receive the flow at the relative angle β_b' and relative Mach number M_b', turning it to β_c' and diffusing it to M_c'. The stator receives the flow at angle β_c and Mach number M_c, turning it to β_d and diffusing to M_d.

If this stage is to be followed by another stage, it is convenient to choose the blading so that $M_d = M_b$ and $\beta_d = \beta_b$. Indeed, in low pressure ratio compressors and pumps the successive stages can be identical. In high pressure ratio compressors some change is necessary from stage to stage because the blade Mach number decreases as the air temperature rises through the successive stages.

The rotor and stator diagrams can be combined into a composite diagram,

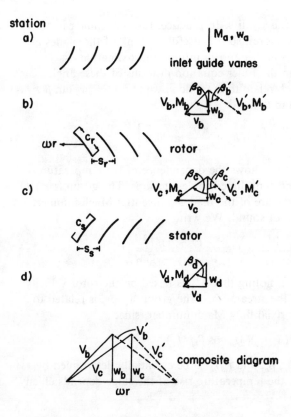

5.3 Development of a stage velocity diagram, The top shows the changes across inlet guide vanes, rotor and stator separately; the bottom shows the composite rotor-stator diagram.

as at the bottom of figure 5.3, which includes all the information for the stage. It also shows the relationships between the velocity changes in the rotor and stator. In particular, the turning introduced by the inlet guide vanes allows the rotor and stator to be nearly reflections of each other about the axial direction, with the same flow deflection, and $\beta_b' \approx \beta_c$, $\beta_c' \approx \beta_d$. It also gives a lower value of M_b' for any particular blade speed ωr.

Characteristic of a closely spaced cascade of blades is that the flow angle leaving the blade row (β_c' for the rotor and β_d for the stator) is closely related to the angle of the trailing edges of the blades. In fact, if the spaces between the blades form long, slender passages, the flow angle at the exit will be controlled by the geometry of the blades at the exit and will be independent of the inlet angle. Denote the chord of the blades by c and the spacing by s; then the extent to which this situation is approached depends on the ratio $\sigma = c/s$, called the solidity. In practice, turbomachine cascades have high solidities (unity or more), so it is useful to think of the blades as controlling the leaving angles β_b, β_c', and β_d in first approximation.

Expressing the velocities in the Euler equation in terms of these angles and the axial velocities, which relate to mass flow, we find $v_b = w_b \tan \beta_b$, $v_c = w_c \tan \beta_c = \omega r_c - w_c \tan \beta_c'$, and

$$\frac{T_{tc}}{T_{tb}} - 1 = \left[\frac{(\omega r_c)^2}{c_p T_{tb}}\right]\left[1 - \left(\frac{w_c}{\omega r_c}\right)\left(\tan \beta_c' + \frac{w_b r_b}{w_c r_c}\tan \beta_b\right)\right]. \tag{5.9}$$

In this form, the Euler equation shows the dependence of the temperature ratio of the stage on the important dimensionless groups. The group $(\omega r_c)^2/c_p T_{tb}$ is proportional to the square of the blades' tangential Mach number, based on the upstream speed of sound. We write

$$\frac{(\omega r_c)^2}{c_p T_{tb}} = \left(\frac{(\omega r_c)^2}{\gamma R T_b}\right)\left(\frac{\gamma - 1}{1 + \frac{1}{2}(\gamma - 1)M_b^2}\right) = \frac{(\gamma - 1)M_T^2}{1 + \frac{1}{2}(\gamma - 1)M_b^2}.$$

to emphasize this dependence, noting that M_T is based on the rotor velocity at the *downstream* radius of the streamtube. The group $w_c/\omega r_c$ is related to the mass flow rate, or to the axial flow Mach number, since

$$w_c/\omega r_c = (w_c/w_b)(w_b/\omega r_c) = (w_c/w_b)(M_b \cos \beta_b/M_T).$$

Thus, if we take w_c/w_b, β_b, β_c', β_d, and r_c/r_b to be essentially determined by the *geometry* of the blading, the temperature rise of the stage depends chiefly on M_T and M_b/M_T,

$$\frac{T_{tc}}{T_{tb}} - 1 = \left[\frac{(\gamma - 1)M_T^2}{1 + \frac{1}{2}(\gamma - 1)M_b^2}\right]\left[1 - \frac{w_c}{w_b}\cos \beta_b \frac{M_b}{M_T}\left(\tan \beta_c' + \frac{w_b r_b}{w_c r_c}\tan \beta_b\right)\right].$$

$$\tag{5.10}$$

From this we deduce that the temperature rise is increased by (1) increasing blade Mach number, (2) decreasing M_b, (3) decreasing β_b and β_c'. Each of these changes degrades some other characteristic of the compressor or leads to a problem. As the blade Mach number is raised above the sonic value, shock losses on the blades increase the entropy rise in the stage, tending to reduce the efficiency. The advantage of high M_T is such that many modern compressors operate with $M_b' > 1$, in spite of these losses. The magnitude of M_b controls the mass flow of the compressor, per unit of area, so it is desirable to increase M_b to as near unity as other limitations will allow. The practical range for aircraft engine compressors is from 0.5 to 0.8. From the velocity triangles, it will be clear that reducing β_c' implies that the rotor must turn the flow through a large angle. This would imply increased losses in the rotor unless the solidity σ_r is very large. Reducing $\beta_b \approx \beta_d$ implies a similar situation in the stator and in addition increases M_b'.

To quantify this complex situation and show how a near-optimum stage design is arrived at for a given set of requirements, we must relate the losses in the blade row to the flow angles. If the losses are assumed known in the sense that the efficiency defined by (3.1) is known, then (5.10) yields the pressure ratio of the compressor stage.

5.2 Compressor Geometry and the Flow Pattern

To determine the temperature rise of an existing compressor stage the flow Mach numbers and angles contained in (5.10) must be related to the geometry, rotative speed, and any other relevant characteristics of the machine. The problem of turbomachine fluid mechanics posed in this way is called the "direct" problem. It can also be posed in the opposite sense; namely, given the desired flow angles and Mach numbers, determine the required geometry. This is the "inverse" problem faced in design. In this discussion the inverse problem will be considered in the main because it is the best vehicle for exhibiting the limitations and design compromises that lead to the prominent characteristics of turbomachinery. The direct problem will be discussed in a qualitative way to bring out the main features of off-design behavior.

The indirect or design problem is conventionally divided into two parts. One imagines first an axisymmetric (no variation in the tangential direction) "throughflow" with axial, tangential, and radial components of velocity; all of these can change discontinuously (or, in more refined analyses, gradually) at axial locations corresponding to the blade rows. In this limit, the blade rows are considered thin "actuator discs" which change the momentum and energy of the fluid, but the variations from blade to blade in the tangential direction are neglected. The throughflow is to a certain approximation determined by the same changes in velocity and angle that enter the Euler equation; thus the throughflow problem can be addressed without reference to the details of blade shape once the velocity triangles are set.

The second part of the problem is to determine the blade shapes required to produce the assumed velocity structure. At present it is usually approached by idealizing the blade row as an infinite rectilinear cascade, as we did in section 5.1.2, and much dependence is put on empirical cascade data.

5.2.1 The Axisymmetric Throughflow

As the flow passes through the "actuator disc," its tangential velocity is changed; its pressure and to a lesser extent the other velocity components are also changed. The change in tangential velocity will generally cause an unbalanced centrifugal force, which leads to a radial acceleration of the fluid. If we focus on a streamtube, as in figure 5.2, this means that the streamtube tends to shift radially as it passes through the blade row. This is the effect, modified by the boundary conditions imposed by the cylindrical hub and casing, that we wish to consider. One might expect that the radial accelerations induced would die out far enough upstream and downstream of the blade row and that the streamtube would assume a "radial equilibrium" position, which is different downstream than upstream. We consider here only this "radial equilibrium" limit, as it shows the physical effects most simply.

Suppose that the equilibrium radii of the streamtube upstream and downstream are r_b and r_c; then the Euler equation (5.7) can be expanded to

$$c_p T_c + \tfrac{1}{2}(u_c^2 + v_c^2 + w_c^2) - c_p T_b - \tfrac{1}{2}(u_b^2 + v_b^2 + w_b^2) = \omega(r_c v_c - r_b v_b). \quad (5.11)$$

The centrifugal force on each fluid element must be offset by the radial pressure gradient, so $dp/dr = \rho v^2/r$ at each station. Suppose that a polytropic exponent can be defined so that $p \propto T^{n/(n-1)}$. If the flow along the streamtube is isentropic $n = \gamma$, but if the entropy increases $n < \gamma$. With this relation,

$$(1/T)(dT/dr) = [(n-1)/n](1/p)(dp/dr) = [(n-1)/n](v^2/rRT)$$

or

$$c_p(dT/dr) = [\gamma/(\gamma-1)]R(dT/dr) = [\gamma/(\gamma-1)][(n-1)/n](v^2/r) \equiv \mu(v^2/r)$$

where $\mu = 1$ for isentropic flow.

Differentiating (5.11) with respect to r, substituting for dT/dr, and assuming that $u^2 \ll (v^2 + w^2)$, we find

$$\begin{aligned}
w_c\left(\frac{dw_c}{dr_c}\right) - w_b\left(\frac{dw_b}{dr_b}\right) &= (\mu - 1)\left[\frac{v_b^2}{r_b} - \frac{v_c^2}{r_c}\right] \\
&+ \left[\frac{v_b}{r_b} - \omega\right]\left[\frac{d(v_b r_b)}{dr_b}\right] - \left[\frac{v_c}{r_c} - \omega\right]\left[\frac{d(v_c r_c)}{dr_c}\right].
\end{aligned} \quad (5.12)$$

Prescribing the inlet and outlet tangential velocity distributions in r thus allows the calculation of the change in axial velocity as a function of r. In the following μ will be put to unity, corresponding to isentropic flow, to simplify the argument.

Suppose first that the flows both upstream and downstream are potential vortices, so that $v_b r_b = B_b w_a r_T$ and $v_c r_c = B_c w_a r_T$, where B_b and B_c are constants, w_a is a reference axial velocity, and r_T is the tip radius of the blade row. It follows immediately that $d(v_b r_b)/dr_b = 0$ and $d(v_c r_c)/dr_c = 0$, so

$$w_c(dw_c/dr_c) - w_b(dw_b/dr_b) = 0, \qquad v_b r_b = B_b w_a r_T, \quad (5.13)$$

or

$$w_c^2 - w_b^2 = \text{const}, \qquad v_c r_c = B_c w_a r_T,$$

so the axial velocity distribution is changed only by a constant when a potential vortex increment is added by a blade row to an already existing potential vortex. Notice that whether the blade row is rotating or not is irrelevant in the determination of w_c.

To determine the temperature rise in the blade row, the assumed forms for v_b and v_c are substituted in the Euler equation (5.7) to give

$$c_p(T_{tc} - T_{tb}) = (B_c - B_b)w_a \omega r_T. \tag{5.14}$$

The important result is that $(T_{tc} - T_{tb})$ is independent of r. This situation or a generalization of it is usually desirable in a compressor. If the blade row produced a much higher pressure ratio at one radius than another, the portion of the annulus with higher pressure ratio might tend to pump fluid backward through the portion with lower pressure ratio. If losses vary with radius, the equal pressure rise condition implies some variation of $T_{tc} - T_{tb}$ with r, but since we are assuming isentropic flow, uniformity of pressure ratio implies uniformity of temperature ratio.

Because of the design simplicity engendered by uniform w, some early aircraft engine compressors were of vortex design; the inlet guide vanes produced the flow $v_b r_b = B_b w_a r_T$, and the rotor modified it to $B_c w_a r_T$. Indeed, such blading is still used in some stationary gas turbines where weight is not a problem and where there is no great premium on mass flow per unit of frontal area, hence on low hub/tip radius ratios. It has the disadvantage, however, that the tangential velocity at the rotor inlet is largest at the hub, where the blade velocity is least, and smallest at the tip, where the blade velocity is greatest. Thus, the rotor velocity triangles at hub and tip are as shown at the top of figure 5.4 for a blade row with (hub/tip radius ratio) of 0.5. Note that a very large turning is required in the rotor at the hub and that the rotor blades are highly twisted. Also the swirl introduced by the inlet guide vanes reduces M_b' less at the tip, where the value of M_b' is largest, than at the root. This type of guide vane is then not very effective for reducing the rotor tip relative Mach number and the associated shock losses.

For these reasons inlet guide vanes that produce a more nearly solid body rotation are better. In this case, for the inlet guide vanes we would have

$$v_a = 0, \qquad v_b = A w_a(r_b/r_T) \tag{5.15}$$

where station a is ahead of the guide vanes and b behind them, as in figure 5.3. From 5.12,

$$w_b(dw_b/dr_b) = -(A w_a/r_T)d[A w_a(r_b^2/r_T)]/dr_b$$

and integrating gives

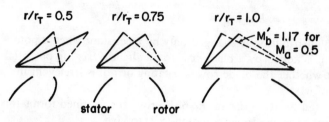

a) Free-vortex inlet guide vanes; $B_b = 0.5$, $B_c = 1.0$

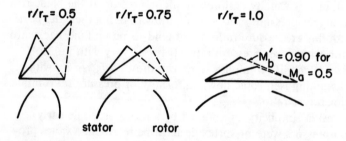

b) Solid-body inlet guide vanes; $A = 1$, $B = 0.5$

5.4 Velocity triangles at hub, mid-span, and tip, for (a) free-vortex inlet guide vanes and (b) solid-body inlet guide vanes.

$$(w_b/w_a)^2 = -2A^2(r_b/r_T)^2 + \text{const.} \tag{5.16}$$

The constant of integration is evaluated by noting that the mass flow through stations a and b must be the same. This condition can be written

$$\int_{r_{Ha}}^{r_{Ta}} \rho_a w_a r \, dr = \int_{r_{Hb}}^{r_{Tb}} \rho_b w_b r \, dr. \tag{5.17}$$

To this point the flow has been treated as compressible and isentropic. To proceed further in this way, we must determine $\rho_b(r)$, and this introduces the pressure ratio p_b/p_a or the Mach number M_b. Consistency also requires that $r_{Tb} \neq r_{Ta}$ and $r_{Hb} \neq r_{Ha}$. The analysis can be carried through in this way, but for the sake of simplicity we will assume here that $\rho_b = \rho_a = \text{const.}$ It is then reasonable to assume also that $r_{Tb} = r_{Ta} = r_T$ and $r_{Hb} = r_{Ha} = r_H$. This incompressible limit, valid in fact only for low Mach numbers, gives considerable insight into the flow patterns while greatly simplifying the calculations. A further simplification can be had by noting that w_b/w_a does not differ greatly from unity for many cases of interest, so we can write

$(w_b/w_a) = \{1 - 2A^2[(r_b/r_T)^2 + \text{const.}]\}^{1/2} \approx 1 - A^2[(r_b/r_T)^2 + \text{const.}].$

Substituting this in (5.17), and evaluating the constant leads to

$$(w_b - w_a)/w_a = A^2(\{[1 + (r_H/r_T)^2]/2\} - (r_b/r_T)^2) \qquad (5.18)$$

so that w_b varies parabolically, decreasing from hub to tip as shown in figure 5.5 for three values of r_H/r_T. The resultant guide vane outlet (rotor inlet) velocities are shown at the bottom of figure 5.4 for $A = 1.0$, $r_H/r_T = 0.5$ which give the same flow angles at the mid radius as in the vortex flow at the top of the figure. For the same ωr_T the tip relative Mach number M_b' is considerably smaller (0.90 compared to 1.17 for $M_a = 0.5$).

To provide a radially constant stagnation temperature increment, a rotor behind this guide vane row must introduce a free vortex velocity increment;

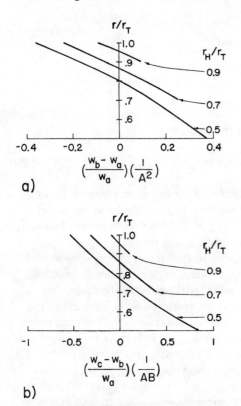

a)

b)

5.5 Axial velocity increments due to blade rows. (a) Axial velocity increment from inlet guide vanes producing solid-body rotation, as function of hub/tip radius ratio and radius. See equation (5.15). (b) Axial velocity increment due to rotor (B positive) or stator (B negative) operating in solid-body flow, but adding vortex increment. See equation (5.19).

so suppose that

$$v_c = Aw_a(r_c/r_T) + Bw_a(r_T/r_c);$$ (5.19)

then

$$w_c(dw_c/dr_c) - w_b(dw_b/dr_b) = -(Bw_ar_T/r^2)d(Aw_ar^2/r_T)/dr.$$

Integrating gives

$$\tfrac{1}{2}w_c^2 - \tfrac{1}{2}w_b^2 = -2ABw_a^2 \log (r/r_T) + \text{const.}$$

To be consistent with our assumption that $(w_b - w_a)/w_a \ll 1$, we again take $(w_c - w_b)/w_a \ll 1$ and write

$$(w_c^2 - w_b^2)/w_a^2 = (w_c - w_b)(w_c + w_b)/w_a^2 \approx 2(w_c - w_b)/w_a.$$

Thus

$$(w_c - w_b)/w_a \approx -2AB \log (r/r_T) + \text{const.}$$

Continuity of the total mass flow requires that

$$\int_{r_H}^{r_T} [(w_c - w_b)/w_a]r \, dr = 0,$$

and this determines the constant with the result that

$$\frac{w_c - w_b}{w_a} = AB\left[\frac{2 \log (r_T/r_H)}{1 - (r_H/r_T)^2} - 1 - 2 \log (r/r_H)\right].$$ (5.20)

This velocity increment is shown in figure 5.5 for the same three values of r_H/r_T chosen for $(w_b - w_a)/w_a$. For B positive, as it would be in the rotor, the trend is similar to that of $(w_b - w_a)/w_a$, so that the rotor further increases the axial velocity at the hub and decreases it at the tip.

The same expression (5.20) is applicable to the stator, taking w_b as w_c and w_c as w_d, if B is simply assigned the appropriate negative value. In particular, if $B_s = -B_r$ so that the stator removes the swirl put in by the rotor, then $(w_d - w_c) = -(w_c - w_b)$ and the flow at the stator outlet is identical to that at the rotor inlet. It is then possible to put several stages, one after the other, without the axial velocity perturbations due to the blade rows becoming large.

At the bottom of figure 5.4 the rotor exit velocities are shown, as determined by (5.20) for $B = 0.50$, which gives the same velocity triangle at mid radius and the same temperature rise as for the free vortex case.

Several points should be noted. First, since these velocity triangles show the radial equilibrium conditions far upstream and downstream of each blade row, they exaggerate the axial velocity changes somewhat. The changes actually take place over an axial distance of the order of the height of the flow annulus $r_T - r_H$, with half of the change upstream of the center of the blade row and half downstream. A detailed exposition of this variation may be found in reference 5.1. Second, we note that the flow deflection in the rotor at the tip is small; in fact, it is slightly negative, away from the axis. The rotor does positive work nevertheless, because the decrease in velocity through it results in a positive tangential velocity increment. The turning at the rotor hub is large, but not so large as for the vortex flow.

Before proceeding to a discussion of the blading, we estimate the pressure ratio this compressor stage would produce. From (5.14),

$$\frac{T_{tc}}{T_{tb}} - 1 = \left[\frac{(\gamma - 1)M_a}{1 + \frac{1}{2}(\gamma - 1)M_a^2}\right]\left[\frac{M_T}{(1 + \frac{1}{2}(\gamma - 1)M_b^2)^{1/2}}\right]B.$$

If we limit the rotative speed to a value that gives a tip relative Mach number M_b' of 0.90 to avoid shock losses, then for the solid body inlet guide vanes $M_T = 1.33$ for $M_a = 0.5$, and $(T_{tc}/T_{tb}) - 1 = 0.253$ corresponding for isentropic flow to $\pi = p_{tc}/p_{tb} = 2.20$. Limiting the vortex design similarly to $M_b' = 0.90$ gives $M_a = 0.384$, $M_T = 1.02$, and $(T_{tc}/T_{tb}) - 1 = 0.149$ or $p_{tc}/p_{tb} = 1.63$. These are high pressure ratios for blading operating at such Mach numbers.

The preceding simple technique for estimating the throughflow should be regarded only as schematic. In the actual design of modern turbomachinery, numerical techniques of flow computation are used to develop the three-dimensional axisymmetric flow taking account of radial accelerations, compressibility, and complex hub and casing shapes. The blades are represented by force distributions and losses are accounted for. A description of such techniques can be found in reference 5.2.

5.2.2 Subsonic Blading

The usual practice in determining the blade shapes required to give the flow angles, determined from the axisymmetric throughflow, is to treat the blades as elements of an infinite two-dimensional cascade whose blades lie perpendicular to the stream surfaces of the throughflow. Experimental data from tests of families of two-dimensional cascades is used to determine the blade shape for a series of stream surfaces at different radii, and the complete blade shape is then determined by interpolating between these radii.

The notation describing such a cascade is shown in figure 5.6. In addition to the items shown there, a complete description of the cascade requires specification of the shape of the camber line, which might be either a circular arc or a parabola, tangent to the directions K_1 and K_2, and specification of the thickness distribution, or some other specification of the shape of the airfoil surface. For supersonic blading a series of circular arcs is used to define the surface.

Given the geometry of the cascade, the flow is completely specified for sub-

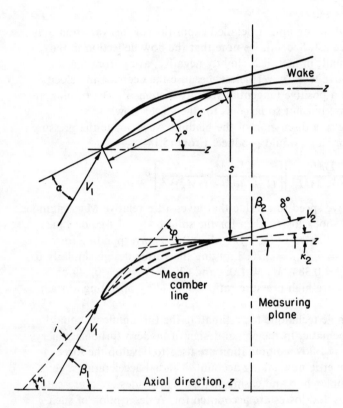

5.6   Notation for cascade.

sonic flow if the incidence i and the upstream Mach number are set. For
fully supersonic approach velocities, it is also necessary to specify the downstream pressure, in ratio to that upstream, in order to locate shock positions.
The performance can then be presented in a number of ways, but perhaps the
most useful is to give the total pressure loss coefficient $\bar{\omega}_1$ and the deviation
angle $\delta°$. The loss coefficient is defined as

$$\bar{\omega}_1 = \Delta p_t/(p_{t1} - p_1) = (p_{t1} - \bar{p}_{t2})/(p_{t1} - p_1) \qquad (5.21)$$

where $\bar{p}_{t2}$ is a mass average over the flow downstream of the cascade. The
deviation is the difference between the actual downstream flow angle β_2 and
the angle of the tangent to the mean camber line at the trailing edge. Variations of loss with incidence and Mach number are given in figure 5.7 for four
types of blading. For each value of M_1 there is an incidence that gives
minimum $\bar{\omega}_1$, and the loss rises much more rapidly with deviation from this
"minimum loss incidence" as M_1 increases. From this one would expect the

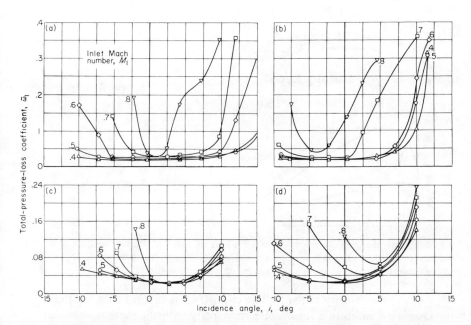

5.7 Variations of loss factor with incidence and inlet Mach number for four types of blading: (a) C4 circular arc blade, $\phi = 25°$, maximum thickness/chord $= 0.1$, $\sigma = 1.333$, $\gamma = 37.6°$. (b) C4 parabolic arc blade, $\phi = 25°$, thickness/chord $= 0.1$, $\sigma = 1.333$, $\gamma = 37.6°$. (c) Double-circular arc blade, $\phi = 25°$, thickness/chord $= 0.105$, $\sigma = 1.333$, $\gamma = 42.5°$. (d) Sharp-nosed blade, $\phi = 27.5°$, thickness/chord $= 0.08$, $\sigma = 1.15$, $\gamma = 30°$. (From reference 5.3.)

deterioration of compressor performance with deviations from the design point to be more rapid as the Mach number is increased.

A minimum-loss incidence angle is defined as the average of the values that give loss twice the minimum value. Figure 5.7 shows that the value $\bar{\omega}_{min}$ of loss at this incidence varies with M_1 and blade design, but weakly compared to the variations of the $\bar{\omega}$ versus i curves. This loss occurs when the flow angles minimize the adverse pressure gradients, and consequent boundary layer thickening, on the blade surfaces.

A semiempirical analysis of cascade data has led to the observation that the minimum loss factor $\bar{\omega}_{min}$ can be correlated in terms of a "diffusion factor" D defined as (reference 5.3)

$$D = 1 - (V_2/V_1) + |v_2 - v_1|/2\sigma V_1. \qquad (5.22)$$

The reasoning is as follows. An airfoil has velocity distributions on suction and pressure surfaces as indicated schematically in figure 5.8. Most of the boundary layer growth and hence most of the wake thickness occur because of the diffusion of the flow along the suction surface from the peak velocity

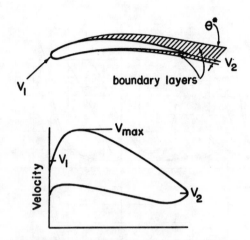

5.8 Schematic diagram of velocity distributions on suction and pressure surfaces of a blade, showing the diffusion from a maximum velocity V_{max} to the final velocity V_2, and the resultant thickening of the boundary layer on the suction surface.

V_{max} to V_2, the velocity at the trailing edge. If θ^* is the wake momentum thickness (the width of a region of zero velocity with the same momentum defect as the wake), one would expect θ^*/c to depend on some function of $(V_{max} - V_2)/V_1$. To relate $(V_{max} - V_2)/V_1$ to the flow angles we note that the lift in first approximation is $L \approx (\rho c/2)(V_{max}^2 - V_2^2)$ and that the tangential velocity change in the blade row is related to the lift by

$$\rho V_1(v_2 - v_1)s \sim L \sim (\rho c/2)(V_{max}^2 - V_2^2).$$

If $(V_{max} - V_1) \ll V_1$ and $(V_1 - V_2) \ll V_1$, then

$$(v_2 - v_1)/V_1 \sim (c/s)[(V_{max} - V_2)(V_{max} + V_2)]/2V_1^2 \sim \sigma(V_{max} - V_2)/V_1$$

so that $(V_{max} - V_2)/V_1 \sim (v_2 - v_1)/\sigma V_1$. The simplest dependence is a linear one, hence the last term of D. The first term follows simply from the diffusion of the flow from V_1 to the lower velocity V_2 according to Bernoulli's equation. Experimental evidence that θ^*/c actually correlates with D is given in figure 5.9.

To convert from θ^*/c to the loss parameter $\bar{\omega}_1$, imagine a wake of zero velocity and width θ^* as in figure 5.10. The (geometrical) average of downstream stagnation pressure is

$$\bar{p}_{t2} = [p_{t1}(s \cos \beta_2 - \theta^*) + p_2 \theta^*]/(s \cos \beta_2);$$

the first term represents the inviscid flow and the second the wake. Substituting in (5.21) gives

$$\bar{\omega}_1 = (\theta^*/s \cos \beta_2)(p_{t1} - p_2)/(p_{t1} - p_1),$$

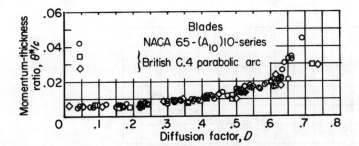

5.9 Correlation of wake momentum thickness that θ^* as fraction of blade chord, with diffusion factor D at minimum loss incidence and $Re = 2 \times 10^5$ (from reference 5.3).

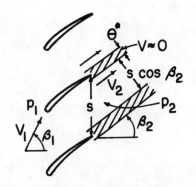

5.10 Schematic showing conversion from wake momentum thickness θ^* to the loss factor $\bar{\omega}_1$.

and since

$$p_{t1} - p_1 = \rho V_1^2/2 = \rho w_1^2/2 \cos^2 \beta_1, \, p_{t1} - p_2 \approx \rho V_2^2/2 \approx \rho w_1^2/2 \cos^2 \beta_2,$$

we get

$$\bar{\omega}_1 = (\cos \beta_1/\cos \beta_2)^2 (\sigma/\cos \beta_2)(\theta^*/c).$$

Hence the proper correlation is of the form

$$\bar{\omega}_1 (\cos \beta_2/\sigma)(\cos \beta_2/\cos \beta_1)^2 = f(D).$$

Such a correlation is shown in figure 5.11. The loss rises rapidly above $D = 0.6$, indicating separation.

For design purposes we also need the deviation δ°. According to Carter's rule

$$\delta^\circ = (m_c/\sqrt{\sigma})\phi \tag{5.23}$$

where m_c depends on the blade chord angle γ° and is given in figure 5.12.

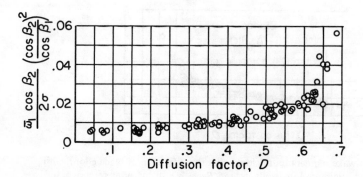

5.11 Correlation of loss parameter with diffusion factor D (from reference 5.3).

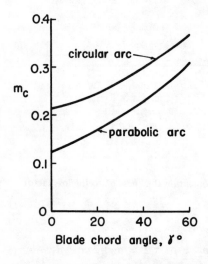

5.12 Parameter for use in Carter's rule for deviation, equation (5.23).

These data, together with figure 5.11, allow us to estimate the exit flow angle and loss for the type of subsonic blade sections for which the correlations were derived.

None of the correlations, however, expresses any effect of Reynolds number. Yet we know that because the loss is a viscous effect, it must depend on Reynolds number. In fact, the correlations are valid only above some limiting Reynolds number, in the range of 2×10^5. Variations of $\bar{\omega}_1$ and $\delta°$ for Reynolds numbers from 10^4 to 3×10^5 are shown in figure 5.13 for a series of incidences. Note that the loss factor and deviation both drop abruptly around 10^5, then remain nearly constant for larger Reynolds numbers. For a compressor blade rotating at a Mach number of unity in air at

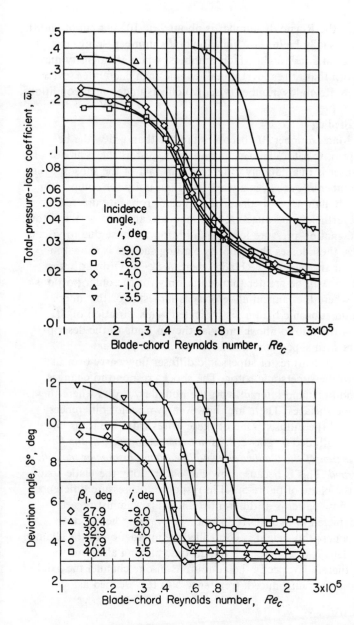

5.13 Variations of loss parameter and deviation with Reynolds number for various incidences (from reference 5.3).

standard conditions, the Reynolds number is about 2×10^5 per cm of chord, so that blading will usually be in the "high Reynolds number" range except at very high altitudes and low flight Mach numbers, where the density may be less than one-tenth that at sea level. Small engines can encounter severe problems due to low Reynolds number (viscous) effects under such conditions.

5.2.3 Supersonic Blading

When the relative Mach number M_b' becomes greater than unity, the behavior of the compressor rotor changes qualitatively, just as the behavior of the supersonic diffuser is qualitatively different from that of the subsonic one. The axial Mach number $M_b' \cos \beta_b'$ is always less than unity for compressor blading; thus even though $M_b' > 1$, disturbances can propagate upstream from the rotor. Nevertheless, because $M_b' > 1$, shocks can form on the rotor blading, causing stagnation pressure losses in addition to those due to viscous effects. To minimize these losses sharp-nosed blading such as is shown in figure 5.14 is used for the part of the blade where $M_b' > 1$. Ordinarily the relative Mach number will be greater than unity only for the outer portions of the rotor blade where the tangential velocity is largest. For the inner portions $M_b' < 1$ and subsonic blading is indicated. This variation of M_b' from below unity at the hub to above unity at the tip leads to the description of compressors with supersonic tips as *transonic compressors*.

Many of the physical features of supersonic diffuser flow carry over at least qualitatively to the supersonic rotor. The shocks respond to changes in downstream conditions and interact with the boundary layers where they impinge on the blade surfaces. There are, however, two major differences between the rotor and the diffuser. One of these arises from the transonic character of the flow alluded to above. *It is not correct to think of the flow in the supersonic portion of the blading as two-dimensional insofar as mass flow continuity is concerned.* This is because flow can occur along the blade span (perpendicular to the plane of figure 5.14) in response to pressure differences set up by different flow patterns at different radii. Thus we must regard the flow channel depicted in figure 5.14 as one with varying height perpendicular to the paper; this height is controlled by interaction of the channels at different radii. (This is of course true of subsonic blading also, but the effects are weaker there.) To specify the conditions in one channel the axial velocity density ratio Ω is introduced. (reference 5.4) It is defined as

$$\Omega = (\rho_b V_b' \cos \beta_b')/(\rho_c V_c' \cos \beta_c')$$

and is nothing more than the ratio of mass flows per unit area at the upstream and downstream stations. If the flow were two dimensional, Ω would be unity. If $\Omega < 1$ the channel is constricted downstream, corresponding to

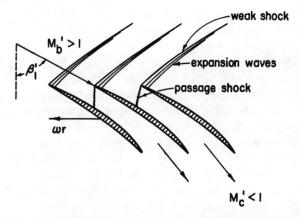

5.14   Blading for supersonic relative velocity, showing incident flow aligned with suction surface of blades, and weak shocks and expansions running upstream.

closing the nozzle on our schematic diffuser of figure 4.15, and if $\Omega > 1$ the channel expands downstream. These variations have the same qualitative effects on the shock in the rotor that they do on the shock in the diffuser.

The second major difference between rotor and diffuser is that the *shocks propagating from the blades' leading edges must be weak* far upstream of the blades. To see this assume first that a blade does generate a strong oblique shock off its suction (low-pressure) side. By symmetry every other blade must do likewise with the result that a series of shocks would run upstream. In the ideal (two-dimensional) case they would proceed far upstream; thus the flow would pass through a large number of such shocks, being turned in the same direction in each. This is impossible because it would result in very large turning; hence the shocks must be weak far upstream and must in fact alternate with expansions that cancel their turning. It follows that the flow cannot be turned by the suction side of the supersonic portion of the blading. It must be aligned with the suction side of the blade, so that *the supersonic rotor operates with constant β_b'*, aligned with the suction surface of the blade. This argument is two dimensional and must be qualified by the observation that transonic three-dimensional effects may cause variations in shock strength as they propagate upstream. Experiments tend to confirm its validity, however.

Suppose now that we know the value of Ω or can control it so as to position the shock at the most favorable location; it has been proposed that the stagnation pressure loss be estimated as the sum of that due to the shock and that obtained from the diffusion factor correlation of subsonic blading, discussed in section 5.2.2. The exact position and strength of the shock are difficult to predict because of the effect of varying Ω. As a first approximation, it can be assumed to be a normal shock at M'_b. For want of better information the deviations given by the correlation can be used provided $M'_c < 1$, the usual situation at least at the design point.

Early attempts to realize the potential of supersonic blading were disappointing; the efficiencies were generally well below those predicted by methods not very different from this. These early experiments used blading of high hub/tip radius ratio, with supersonic conditions over the entire blade height. Later blading was designed with lower hub/tip ratios, hence *transonic* in the sense that the hub sections were subsonic even though the tips were supersonic. This blading gave much better efficiencies, and this type of blading has been exploited in modern compressor designs. Although the reasons for the success of transonic compressors are not entirely clear, a major factor appears to be the existence of both subsonic and supersonic regions. The presence of the subsonic regions allows the flow to adjust radially, so that the extreme pressure rises and choking effects associated with the supersonic tips are moderated by the subsonic regions.

It has been demonstrated, for example, that in linearized theory there is no sudden, disastrous drag rise when the blade relative Mach number at the tip passes through unity, in contrast to the situation in two-dimensional linearized airfoil theory (reference 5.5).

Design techniques for supersonic blading are a subject of intensive study as of this writing, and the serious reader is advised to consult the literature for new developments.

5.2.4 The Loss Factor and Efficiency

To determine the relationship between the loss factor $\bar{\omega}_1$ and compressor efficiency, it is helpful first to establish the connection between $\bar{\omega}_1$ and the entropy rise. The reason for this is that the entropy, being a property of the fluid, is independent of the transformation from rotor (moving) coordinates to stator (stationary) coordinates. This is not true of the stagnation pressure, which depends on the coordinate system to which the velocity is referred.

The entropy is related to the temperature and pressure of the fluid by

$$s - s_0 = c_p \log(T/T_0) - R \log(p/p_0)$$

where 0 denotes some reference state. Since the process connecting the stagnation state to the local thermodynamic state of the gas is isentropic, we can also write

$$s - s_0 = c_p \log(T_t/T_0) - R \log(p_t/p_0)$$

where

$$T_t = T\{1 + \tfrac{1}{2}(\gamma - 1)M^2\} \quad \text{and} \quad p_t = p\{1 + \tfrac{1}{2}(\gamma - 1)M^2\}^{\gamma/(\gamma - 1)}.$$

If a viscous process lowers p_t from p_{t1} to p_{t2}, and if we assume $T_{t2} = T_{t1}$, the change in entropy is then

$$s_2 - s_1 = -R \log (p_{t2}/p_0) + R \log (p_{t1}/p_0) = R \log (p_{t1}/p_{t2}).$$

Now the loss factor for the rotor is defined as

$$\bar{\omega}_b' = (p_{tb}' - p_{tc}')/(p_{tb}' - p_b)$$

where the primes indicate that the values are referred to rotor coordinates. Solving for p_{tb}'/p_{tc}' and substituting gives

$$s_c - s_b = -R \log \{1 - \bar{\omega}_b'[1 - (p_b/p_{tb}')]\}.$$

The assumption that $T_{t2} = T_{t1}$ implies, for the stator cascade that the effects of flow unsteadiness (see section 5.1) and heat conduction in the blades are negligible. The latter assumption is easily verified; the former is more subtle. Its validity remains to be demonstrated. For the rotor cascade, the equivalent of (5.5) is

$$\rho D(h + \tfrac{1}{2}u^2)/Dt + \rho\mathbf{u}\cdot[\boldsymbol{\Omega} \times (\boldsymbol{\Omega} \times \mathbf{r})] = \partial p/\partial t$$

where $\mathbf{r}$ is the radius vector from the origin in the coordinate system rotating at angular velocity $\boldsymbol{\Omega}$ with the rotor. The term $\boldsymbol{\Omega} \times (\boldsymbol{\Omega} \times \mathbf{r})$ is the centrifugal force on the fluid particle due to $\boldsymbol{\Omega}$, so $\mathbf{u}\cdot[\boldsymbol{\Omega} \times (\boldsymbol{\Omega} \times \mathbf{r})]$ is the rate at which work is done on the fluid by its *radial motion* in the centrifugal force field. Thus only if radial velocities are small is $c_p T_t = h + \tfrac{1}{2}u^2$ conserved in the rotor coordinates. The energy addition by the centrifugal field is large in radial flow compressors, but within the cascade approximation it is zero if the flow is assumed to be along surfaces of constant radius.

In the cascade approximation, then, p_{tb}' is the stagnation pressure *referred to the coordinates in which $\bar{\omega}_b'$ is determined*, so if stations b and c are ahead of and behind the rotor, p_{tb}' is the rotor inlet stagnation pressure in rotor coordinates. Similarly, if the loss factor for the stator is $\bar{\omega}_c$, then the entropy rise across the stator is $s_c - s_c = -R \log \{1 - \bar{\omega}_c[1 - (p_c/p_{tc})]\}$ where p_{tc} is referred to the *stator* coordinates. The total entropy rise across the stage will be $(s_c - s_b) + (s_d - s_c)$ or

$$s_d - s_b = -R \log \left(\{1 - \bar{\omega}_b'[1 - (p_b/p_{tb}')]\}\{1 - \bar{\omega}_c[1 - (p_c/p_{tc})]\} \right). \tag{5.24}$$

The stage efficiency is defined as

$$\eta_s = \frac{(p_{td}/p_{tb})^{(\gamma-1)/\gamma} - 1}{T_{td}/T_{tb} - 1} = \frac{(T_{td}/T_{tb}) \exp \{-(s_d - s_b)/c_p\} - 1}{T_{td}/T_{tb} - 1}.$$

Substituting for $s_d - s_b$ and taking $\bar{\omega}_b' \ll 1$ and $\bar{\omega}_c \ll 1$, we then find

$$\eta_s = 1 - \frac{\{(\gamma - 1)/\gamma\}\{\bar{\omega}_b'[1 - p_b/p_{tb}'] + \bar{\omega}_c[1 - p_c/p_{tc}]\}}{\tau_s - 1},\tag{5.25}$$

where $\tau_s = T_{td}/T_{tb}$.

Equation (5.25) shows some effect of the relative Mach numbers on η_s since

$$(p_{tb}'/p_b) = \{1 + \tfrac{1}{2}(\gamma - 1)M_b'^2\}^{\gamma/(\gamma-1)}, \ (p_{tc}/p_c) = \{1 + \tfrac{1}{2}(\gamma - 1)M_c^2\}^{\gamma/(\gamma-1)}\tag{5.26}$$

where M_b' is the Mach number relative to the rotor, and M_c that relative to the stator. But to see the overall effect of Mach number on η_s, we must take account of the variation of τ_s with M. From (5.10), we can write $\tau_s - 1 = cM_T^2$, where c depends on the velocity triangles and may in first approximation be assumed independent of M. Expanding (5.26) in M^2 and substituting in (5.25), we find

$$\eta_s \approx 1 - \tfrac{1}{2}(\gamma - 1)\frac{M_b'^2\bar{\omega}_b' + M_c^2\bar{\omega}_c}{cM_T^2}$$

so if we increase all the Mach numbers proportionately, keeping $\bar{\omega}_b'$ and $\bar{\omega}_c$ constant, η_s does not change. This conclusion applies for low M and is valid only if shock losses are small compared to viscous losses. For high Mach numbers, where p_b/p_{tb}' and p_c/p_{tc} begin to approach zero but $\tau_s - 1$ still increases with M_T, η_s would increase according to (5.25), but then shock losses become important and may reverse the trend.

To illustrate the application of these ideas, let us estimate the efficiency of the stage represented by figure 5.4b. Table 5.1 lists the various quantities involved in the estimate. It has been assumed that $M_a = 0.5$, and the solidities of rotor and stator have been set to unity at the tip radius. The table shows that although both D_r and D_s are within reasonable bounds, the rotor has large predicted losses at the tip and the stator has large losses at the hub. Note also that the stator has $M_c > 1$ at the hub, a situation that would lead to increased losses there. As was noted in section 5.2.1, this stage with a pressure ratio of 2.13 at a tip relative Mach number of only 0.90 is beyond the normal limits of design. It was selected to exaggerate the important effects. Lowering the pressure ratio would lower the losses at hub and tip and M_c at the hub, resulting in an estimate for η_s near 0.98 for the entire annulus. This is not unreasonable as an estimate of efficiency taking account only of the blade profile losses.

5.2.5 Non-Two-Dimensional Losses
There are several recognized sources of loss in an axial compressor in addition to the two-dimensional loss on the blading. Some of these are tip clearance leakage, hub and casing shear, and three-dimensional flow in the

Table 5.1 Numerical values leading to estimates of pressure ratio and efficiency for the stage of figure 5.4, accounting only for cascade losses.

Quantity		Hub	Mid Radius	Tip
Rotor	M'_b	0.71	0.76	0.90
	V_c/V_b	1.087	0.708	0.703
	$(v_c - v_b)/2V_b$	0.326	0.208	0.139
	σ_r	2.0	1.5	1
	D_r	0.076	0.431	0.436
	$(\bar{\omega}'_b \cos \beta'_c/2\sigma)(\cos \beta'_c/\cos \beta'_b)^2$	0.006	0.012	0.012
	$\cos \beta'_b$	0.840	0.603	0.358
	$\cos \beta'_c$	0.995	0.858	0.350
	$\bar{\omega}'_b$	0.017	0.021	0.072
	$\bar{\omega}'_b[1 - (p_b/p'_{tb})]$	0.0048	0.0065	0.0294
Stator	M_c	1.13	0.866	0.785
	V_d/V_c	0.588	0.693	0.768
	$(v_d - v_c)/2V_c$	0.221	0.192	0.159
	σ_s	2.0	1.5	1
	D_s	0.522	0.435	0.312
	$(\bar{\omega}_c \cos \beta_d/2\sigma_s)(\cos \beta_d/\cos \beta_c)^2$	0.018	0.012	0.008
	$\cos \beta_c$	0.73	0.57	0.25
	$\cos \beta_d$	0.93	0.79	0.56
	$\bar{\omega}_c$	0.048	0.0238	0.0057
	$\bar{\omega}_c[1 - (p_c/p_{tc})]$	0.026	0.0094	0.0019
	η_s	0.965	0.982	0.965
	τ_s	1.25	1.25	1.25
	π_s	2.13	2.15	2.13

blade passages. They will be described briefly here; more detailed discussions may be found in reference 5.6.

The tip clearance leakage is driven by the pressure difference between pressure and suction sides of the blades. At the gap between the rotor tip and the casing, a jet is driven by this pressure difference; the jet issues from the suction side with a velocity equal to the main flow there but with a different direction. Thus a part of its kinetic energy is dissipated, leading to an increase in entropy.

The hub and casing boundary layers are complex because they pass alternately between moving and stationary blade rows as they develop along the compressor flow path. Since the axial velocity is lower in the boundary layer than in the inviscid flow, the rotor does more work on the boundary layer, which tends to thin it and retard separation but at the same time increase the shear. In general the thickness of these boundary layers is of the same order as the blade spacing; as they pass through the blade rows, a large-scale three-dimensional flow is set up. This may be seen as follows. A pressure difference between the pressure and suction sides of the blade passage is required to turn the inviscid flow. This same pressure difference is felt by the boundary layer. But because the boundary layer flow has less axial momentum, the pressure gradient turns it through a larger angle, leading to the type of flow sketched in figure 5.15. In addition to the losses caused by the shear flow itself, this flow induces changes in flow direction on the blades, which complicate the design.

Unfortunately, straightforward methods for estimating the reduction in efficiency due to these loss mechanisms cannot be given, so that the efficiency a new compressor design will attain is somewhat uncertain. Even in the best of circumstances an axial flow compressor attains its best efficiency only after a long, expensive process of empirical development.

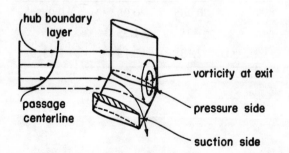

5.15 Schematic of three-dimensional flow caused by boundary layer passing through blade row, showing generation of axial vorticity.

5.3   Limits on Stage Pressure Ratio and the Compromise with Efficiency and Mass Flow

In this section, the information presented in section 5.2 will be assembled to show how the requirements for high mass flow per unit frontal area, high pressure ratio per stage, and high efficiency interact with one another and how they may be traded against each other.

Consider the stage of figure 5.3 with inlet guide vanes. For the sake of simplicity we assume that there are no radial displacements of streamtubes and that the axial velocity is constant throughout. The diffusion factors for rotor D_r and stator D_s are

$$D_r = 1 - (V'_c/V'_b) + [(v_c - v_b)/(2\sigma_r V'_b)],$$

$$D_s = 1 - (V_d/V_c) + [(v_c - v_d)/(2\sigma_s V_c)].$$

From the Euler equation,

$$(T_{tc}/T_{tb}) - 1 = (\omega r/c_p T_{tb})(v_c - v_b) = (\omega r w/c_p T_{tb})(\tan \beta'_b - \tan \beta'_c).$$

Defining a work coefficient ψ as

$$\psi = [(T_{tc}/T_{tb}) - 1]/[(\omega r_T)^2/2c_p T_{tb}] \qquad (5.27)$$

and a dimensionless radius $\zeta = r/r_T$ gives

$$\psi = \zeta(\tan \beta'_b - \tan \beta'_c)(2/\mu) \qquad (5.28)$$

where $\mu = \omega r_T/w$ and $(v_c - v_b)/V'_b = [(\cos \beta'_b \psi)/\zeta](\mu/2)$. If we further assume $v_c - v_d = v_c - v_b$, that is, that the stator takes out the swirl put in by the rotor, then

$$(v_c - v_d)/V_c = [(\cos \beta_c \psi)/\zeta](\mu/2)$$

so that D_r and D_s become

$$D_r = 1 - (\cos \beta'_b/\cos \beta'_c) + (\mu\psi \cos \beta'_b/4\sigma_r\zeta),$$

$$D_s = 1 - (\cos \beta_c/\cos \beta_d) + (\mu\psi \cos \beta_c/4\sigma_s\zeta). \qquad (5.29)$$

For solid-body inlet guide vanes,

$$v_b/w = [v_b(r_T)/w]\zeta \quad \text{and} \quad \tan \beta'_b = [(\omega r_T/w) - v_b(r_T)/w]\zeta \equiv \alpha\zeta.$$

From (5.28), $\tan \beta'_c$ is then given by $\tan \beta'_c = \alpha\zeta - (\mu\psi/2\zeta)$. Computing $\cos \beta'_b$ and $\cos \beta'_c$, taking $\sigma = \sigma_T/\zeta$, and substituting in D_r gives

$$D_r = 1 - \left[\frac{1 + [\alpha\zeta - (\mu\psi/2\zeta)]^2}{1 + (\alpha\zeta)^2}\right]^{1/2} + \frac{\mu\psi}{4\sigma_{rT}[1 + (\alpha\zeta)^2]^{1/2}} \cdot \qquad (5.30)$$

Similarly, $\tan \beta_c = \mu\zeta - \tan \beta'_c = (\mu - \alpha)\zeta + (\mu\psi/2\zeta)$, and if $\beta_d = \beta_b$, then $\tan \beta_d = (\mu - \alpha)\zeta$ so that D_s is

$$D_s = 1 - \left[\frac{1 + (\mu - \alpha)^2\zeta^2}{1 + [(\mu - \alpha)\zeta + \mu\psi/2\zeta]^2}\right]^{1/2} + \frac{\mu\psi}{4\sigma_{sT}\{1 + [(\mu - \alpha)\zeta + \mu\psi/2\zeta)]^2\}^{1/2}} \cdot$$

(5.31)

With the values of D_r and D_s from (5.30) and (5.31), the D-factor correlation gives the loss parameters,

$\lambda_r = (\bar{\omega}_b \cos \beta'_c/2\sigma_r)(\cos \beta'_c/\cos \beta'_b)^2$,

$\lambda_s = (\bar{\omega}_c \cos \beta_d/2\sigma_s)(\cos \beta_d/\cos \beta_c)^2$,

The loss factors are then

$$\bar{\omega}'_b = \lambda_r\left[\frac{2\sigma_{rT}}{\zeta}\right]\left[1 + \left(\alpha\zeta - \frac{\mu\psi}{2\zeta}\right)^2\right]^{1/2}\left[\frac{1 + [\alpha\zeta - \mu\psi/2\zeta]^2}{1 + (\alpha\zeta)^2}\right],$$

(5.32)

$$\bar{\omega}_c = \lambda_s\left[\frac{2\sigma_{sT}}{\zeta}\right]\left[\frac{\{1 + (\mu - \alpha)^2\zeta^2\}^{1/2}\{1 + (\mu - \alpha)^2\zeta^2\}}{1 + [(\mu - \alpha)\zeta + \mu\psi/2\zeta]^2}\right].$$

(5.33)

The relative Mach numbers are also important. Since the axial velocity has been assumed constant, the axial Mach number changes only because of temperature changes; these are small enough to be neglected in an analysis like this, so the relative Mach numbers are

$$M'_b/M_a = \sec \beta'_b = [1 + (\alpha\zeta)^2]^{1/2},$$

(5.34)

$$M_c/M_a = \sec \beta_c = \{1 + [(\mu - \alpha)\zeta + (\mu\psi/2\zeta)]^2\}^{1/2}.$$

(5.35)

If we now assume that the shock losses in the rotor, for $M_b > 1$, are those for a normal shock, they can be found from figure 4.2. The fractional stagnation pressure loss, denoted say $\bar{\omega}'_s$, should be added to $\bar{\omega}'_b$ as given by (5.32) to give the total pressure loss of the rotor.

Finally, then, the stage efficiency can be computed from (5.24) using M'_b and M_c to find p_b/p'_{tb} and p_c/p_{tc}.

The general trend of diffusion factor with radius is shown in figure 5.16 for two stages, both with work coefficients $\psi = 0.5$, and with equal values of $\mu = \omega r_T/w$, one without inlet guide vanes and one with inlet guide vanes. According to (5.27), either would produce a temperature ratio $\tau_s = 1 + (\gamma - 1)M_a^2\mu\psi = 1.1$ or a (lossless) pressure ratio $\pi_s \approx 1.40$. Perhaps most important is that the diffusion factor for the stator D_s increases with decreasing ζ for both designs, so if a maximum acceptable value of D_s is set, it sets a lower limit on ζ and hence on the ratio of hub to tip radius. A similar limit may be set by the rotor if the hub/tip ratio is large, but for small hub/tip ratio an intermediate radius is most critical for diffusion in the rotor.

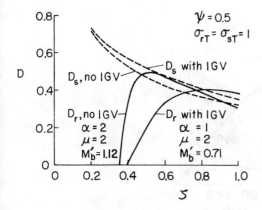

5.16   Variation of diffusion factor of rotor D_r and stator D_s with radius $\zeta = r/r_T$, for stages with and without guide vanes, work coefficient $\psi = 0.5$, and tip solidities $\sigma_{rT} = \sigma_{sT} = 1$.

These two stages, one with and one without inlet guide vanes, operating at the same blade speed $\omega r_T = \mu w_a = 2w_a$ produce the same pressure ratio, but the one with inlet guide vanes has a relative Mach number of only 0.71, compared to 1.12 for the one without inlet guide vanes. The lower Mach number would result in higher efficiency, but the inlet guide vanes can lead to an intense whining noise. For this reason, the design without inlet guide vanes is preferred for large subsonic turbofan engines.

5.3.1   Mass Flow vs Pressure Ratio
In engines for supersonic aircraft there is a great incentive to maximize the mass flow per unit of engine frontal area to minimize drag. This requires both that the axial Mach number M_a be as close to unity as possible and that the useful flow area $\pi(r_T^2 - r_H^2)$ be maximized. The practical limit on M_a is set by the blockage of the flow annulus by the blades. From figure 4.1 we see that if the inlet guide vanes block 16 percent of the annulus, for example, the maximum possible value of M_a is about 0.6; this is typical. With M_a set, the mass flow can be increased only by decreasing r_H/r_T; but for a given limit on D the maximum possible value of ψ then decreases as r_H/r_T decreases with a consequent reduction of the pressure ratio.

By solving (5.30) and (5.31) for ψ, the limiting values can be found for specified D_r and D_s. Some typical results for a stage without inlet guide vanes, $\alpha = \omega r_T/w_a = 2$, and with the maximum diffusion factors in rotor and stator set at 0.5 are shown in figure 5.17. From the lower curves, the stator limits ψ for r_H/r_T above 0.9 and below 0.5, while the rotor limits ψ for the intermediate hub/tip ratios. With these values of ψ_{max}, τ_s follows from (5.27), and if the efficiency is assumed equal to unity π_s is as shown in the upper

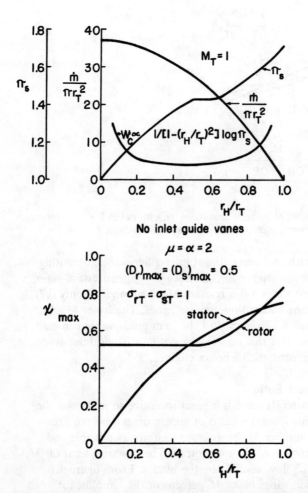

5.17 Variation of maximum work factor ψ_{max} as limited by diffusion factor, with hub/tip radius ratio for a stage without inlet guide vanes, and the corresponding variations of pressure ratio, mass flow per unit frontal area, and compressor weight.

part of figure 5.17. This decrease of π_s with decreasing r_H/r_T is compensated by an increase of mass flow also shown in figure 5.17 for sea level static conditions.

This implies that for a compressor with fixed overall pressure ratio and fixed mass flow there is a choice between compressors of small diameter and many stages on the one hand and compressors of larger diameter and fewer stages on the other. If we assume the compressor weight is proportional to the frontal area and to the number of stages n, $W_c \propto nr_T^2$, then since for a fixed overall pressure ratio $\pi_c = \pi_s^n$ or $n = \log \pi_c/\log \pi_s$, we have $W_c \propto 1/[1 - (r_H/r_T)^2] \log \pi_s$. This function is shown in figure 5.17. For this stage the lightest compressor would have $r_H/r_T \approx 0.45$, but the minimum is very broad, so that variations of r_H/r_T from 0.3 to 0.6 would lead to only small variations in compressor weight.

5.3.2 Pressure Ratio and Efficiency vs Mach Number

Increasing the tangential Mach number of the rotor increases the temperature rise of the stage roughly as M_T^2 if the velocity triangles are kept similar (if $\omega r_T/w_a$ is kept constant), but it is more realistic to think of holding M_a constant while increasing M_T. When this is done, $\alpha = \omega r_T/w_a$ increases and M_b' increases. To see this trend consider a stage of hub/tip ratio near unity in which D_r and D_s are held constant as M_T is increased while M_a is held constant. Solving (5.30) gives ψ as a function of M_T. With this value of ψ the solidity required for the stator to have the specified D_s is found from (5.31). The loss factors $\overline{\omega}_b'$ and $\overline{\omega}_c$ and the relative Mach numbers M_b' and M_c then follow as do the efficiency and pressure ratio.

Such results are shown in figure 5.18 for a stage without inlet guide vanes. The pressure ratio varies from $\pi_s = 1.25$ at $M_T = 0.5$ ($\alpha = 1$) to nearly 4 at $M_T = 2$ ($\alpha = 4$). The efficiency varies from 0.976 at $M_T = 0.8$ to 0.84 at $M_T = 2$; the rapid decrease (increase in $1 - \eta_s$) is due to shock losses in the rotor that occur as M_b' rises above unity, reaching $M_b' \approx 2$ at $M_T = 2$. Although M_c is always well below unity, the stator solidity required to hold D_s to 0.5 becomes as large as 5 at the high rotor speeds, and this would pose a serious problem of flow blockage.

This calculation exaggerates the variation of η_s with $\overline{M}_T$ because the peak value of η_s would be lowered by other effects such as shear on the hub and casing and flow through the clearance gaps at the blade ends. Nevertheless, it does give the right trend and demonstrates that the best blade Mach number depends on the relative importance of efficiency and stage pressure rise (or weight reduction) in a given application. For a long-range cruise application where fuel weight is more important than engine weight, the premium on compressor efficiency would drive M_T toward 1 or perhaps less, while a direct-lift engine which operates for only a few minutes at takeoff

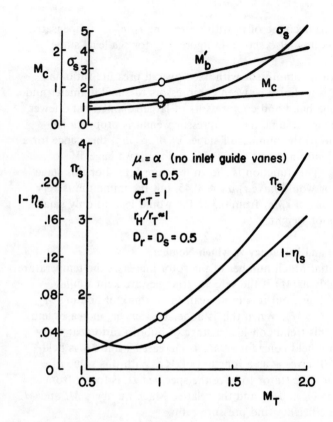

5.18   Variation of pressure ratio and stage efficiency (accounting only for cascade and shock losses) with blade tangential Mach number for a stage without inlet guide vanes, and hub/tip ratio near unity, with corresponding relative Mach numbers and stator solidity.

and landing might optimize at high M_T because the fuel consumption is less important compared to engine weight.

5.3.3   The Effect of Inlet Guide Vanes

Some of the increase in stage pressure ratio with increasing M_T can be had without the penalty due to shock losses by using inlet guide vanes to reduce the rotor relative Mach number as M_T is increased. The effect can be seen by holding α constant while increasing $\mu = \omega r_T / w_a$. This has been done in figure 5.19 where all other parameters are as in figure 5.18. The points marked by circles are identical in these two figures.

We note that $1 - \eta_s$ actually declines slightly with increasing M_T until M_c exceeds unity; then it begins to rise rapidly because of shock losses in the stator (assumed to be those for a normal shock at M_c). For $\pi_s \approx 2.4$ the

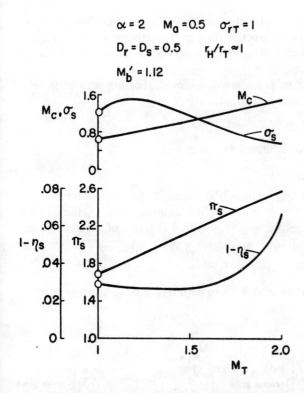

5.19 Effect of inlet guide vanes on the pressure ratio and efficiency of a compressor stage
with fixed rotor relative Mach number, the rotor tangential Mach number increasing with
inlet guide vane turning.

predicted η_s for the stage with guide vanes is 0.96 at $M_T = 1.8$, while for the
one without guide vanes, $\eta_s \approx 0.92$ at $M_T = 1.45$.

Apart from the difference in noise generation the choice between these
two alternatives hinges on the importance of efficiency and the advantages or
disadvantages of high blade speed. The turbine can be a controlling factor
in this choice.

5.4 Stage Performance: Corrected Parameters

From the Euler equation in the form of (5.10) we argued that the temperature
ratio across a stage should depend on the tangential Mach number of the
rotor M_T on the axial flow Mach number M_b (or M_a) and on the flow
geometry as controlled by the blade angles, so we should be able to correlate
say τ_s as a function of M_T and M_a. To obtain the pressure ratio the efficiency

must be introduced, but if η_s is a function only of M_T, M_a, and flow geometry, then $\pi_s = \pi_s(M_a, M_T)$. In practice stage performance can be correlated in this way as long as the Reynolds number is sufficiently large (greater than about 3×10^5 based on blade chord).

A different set of parameters is preferred, although they are equivalent to M_a and M_T. They are the corrected speed, defined as

$$N/\sqrt{\theta} \equiv N/(T_{t2}/T_r)^{1/2} \text{ rpm,}$$

and the corrected weight flow, defined as

$$W\sqrt{\theta}/\delta \equiv W\sqrt{(T_{t2}/T_r)}/(p_{t2}/p_r) \text{ lbs s}^{-1}$$

where T_r and p_r are the sea level standard atmospheric pressure and temperature, 2116.2 lb/ft² and 519°R. That $N/\sqrt{\theta}$ is equivalent to M_T for a given tip diameter is obvious. By expressing W, θ, and δ in terms of the axial Mach number and stagnation conditions, we can show that

$$(W_2\sqrt{\theta_2})/A_2\delta_2 = c(\rho_2 u_2)/(\rho_{t2} a_{t2}) = f(M_2)$$

where for air $c = 85.3$ lb/ft²sec and A_2 is to be interpreted as the flow area at the station 2 where $\rho_2 u_2$ is given.

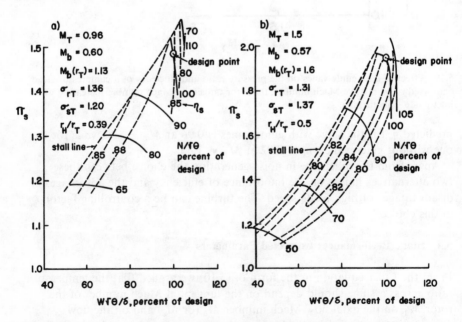

5.20 Performance maps for modern, highly loaded fan stages without inlet guide vanes. (a) With subsonic tangential Mach number (from reference 5.7). (b) With tangential Mach number of 1.5 (from reference 5.8).

The performance of two modern high-performance fan stages is shown in this format in figure 5.20 (references 5.7 and 5.8). They have no inlet guide vanes. One has a low tangential Mach number (0.96) to minimize noise. The other has supersonic tip speed and a considerably larger pressure ratio. Both have high axial Mach numbers.

Several features of these maps should be noted. As the weight flow is reduced at fixed speed, the pressure ratio rises until a limiting value, indicated by the "stall line," where the flow tends to become unsteady. At low corrected speeds the weight flow changes considerably with pressure ratio for both fans. This can be readily understood from (5.10). Because $W\sqrt{\theta}/\delta$ corresponds to M_a, the axial Mach number, increasing $W\sqrt{\theta}/\delta$ lowers τ_s, hence π_s.

But as $N/\sqrt{\theta}$ becomes larger, the constant speed characteristics of both fans become steeper with very little change in $W\sqrt{\theta}/\delta$ as the pressure ratio is changed. In the case of the low tip speed fan this happens because the axial Mach number in the rotor passages is approaching unity as $N/\sqrt{\theta}$ exceeds the design value, so that a drop in downstream pressure does not increase the mass flow. In the case of the high tip speed fan the rotor relative Mach number is greater than unity over a large part of the annulus at design speed and above, and this implies that β'_b is constant, so $W\sqrt{\theta}/\delta$ should depend only on $N/\sqrt{\theta}$.

5.5 Multistage Compressors

High pressure ratio multistage compressors have been developed by placing several stages, each composed of a rotor and stator, in series on the same shaft. The requirements that the successive stages have the same rotative speed and pass the same mass flow lead to some difficult problems at off-design operating conditions. To see how these come about note first that as the pressure builds up through each successive stage, the flow area must be reduced if the axial flow velocity is to remain about the same fraction of blade speed throughout the several stages. The area reduction can be accomplished by increasing the hub radius, decreasing the tip radius, or both. Both have advantages and disadvantages. Decreasing the tip radius lowers the blade tangential Mach number, adding to the decrease due to rising temperature of the air, and so lowers the pressure ratio of the downstream stages. On the other hand, tip clearance losses can be excessive in the last stages if only the hub radius is tapered because the blade heights become small, and stresses in the discs supporting the blades also increase.

In any case the reduction in flow area from front to rear stages, if correct at the design point where the compressor has a large pressure ratio, will not be correct at lower speeds where the pressure ratio is lower. Relative to the

center stages, the front stages will have too large a flow area and the rear
stages too small a flow area; the result is that the axial velocity will be too
low relative to blade speed in the front stages and too high in the rear stages.
This leads to stall of the front stages and "windmilling" of the last ones.
The situation is shown schematically in figure 5.21. The low-speed problem
is serious in compressors with pressure ratios above about 4. Two methods
have been developed for dealing with it. One is to split the compressor into
two independently rotating "spools." At low speeds the front spool then
runs slower relative to the rear spool than at design point, and this eases
the problem in both spools. The second solution to the problem is to use
variable stators, whose angular setting can be adjusted while the engine is
operating, in the first few stages. The first solution was adopted in the Pratt
& Whitney JT-3 turbojet and in the JT3-D turbofan (figure 1.16) which

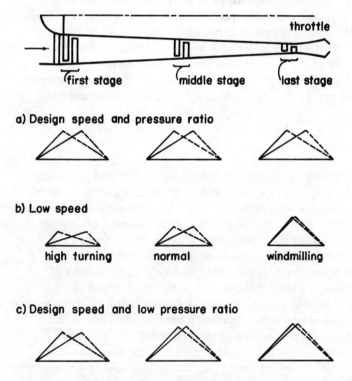

5.21   Off-design behavior of a multistage, high pressure ratio compressor, showing velocity
triangles for (a) design speed and pressure ratio, with similar angles and loading in all
stages (b) low speed (low pressure ratio) operation, with front stages highly loaded and
rear stages windmilling (negatively loaded), and (c) design speed, but low pressure ratio,
with rear stages unloaded but front stage nearly at design condition.

powered most first generation jet transports. The second was adopted in the General Electric J-79 turbojet which powers many military aircraft including the F-4. Both were highly successful. More modern engines such as the P & W F-100 (figure 1.18) and GE CF-6 (figure 1.17) usually incorporate both solutions.

Variations of the axial flow velocity in response to changes in pressure cause the multistage compressor to have quite different mass flow versus pressure ratio characteristics than one of its stages. Suppose that the throttle is opened with the compressor at design speed. The first effect is to lower the pressure and increase the axial velocity in the last stage as shown at the bottom of figure 5.21. This tends to reduce its pressure ratio, and the reduction in pressure ratio causes the next upstream stage to feel a smaller pressure reduction. It in turn reduces the signal and by the time the front stage is reached the reduction in exit pressure may have virtually disappeared. As a result its flow velocity is unchanged, and the compressor mass flow is virtually unchanged by the drop in exit pressure. This leads to very steep constant

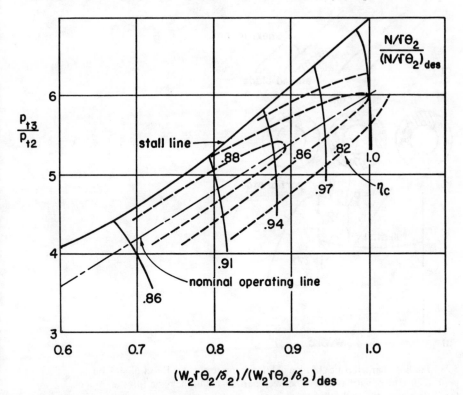

5.22 Performance map for a typical high pressure ratio compressor.

speed lines for high pressure ratio compressors, similar to that of the single transonic stage of figure 5.20. Note, however, that the steepness of the characteristic has quite different origins in the two cases. A typical map for a modern high pressure ratio compressor is shown in figure 5.22.

5.6 Stall and Surge

Stall is critical in compressors because its occurrence limits the pressure ratio available from any given design. From the viewpoint of an individual compressor blade, stall simply implies separation of the flow, usually from the suction surface, but the interactions of any one blade with others and with the overall flow field are so many that the overall behavior of the compressor at stall is complex. For reasons not clearly understood, stall does not occur uniformly around the compressor annulus, but rather in a series of uniformly spaced cells near the hub or near the tip, as sketched in figure 5.23 where the

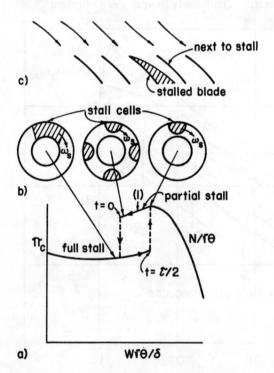

5.23 Stalling characteristics of an axial flow compressor. (a) Effect of stall on performance map, with hysteresis between partial stall branch and full stall branch. (b) Geometry of the stall cells, which rotate at angular velocity ω_s, a fraction of rotor angular velocity. (c) Mechanism for propagation of rotating stall.

shaded areas represent the stall cells. The regions of stall propagate relative to the stalled blade row with a velocity that is a fraction of the tangential flow velocity relative to the stalled blade row; thus in absolute coordinates the cells move in the direction of the rotor at some fraction (usually about 1/2) of rotor speed. This means the cells move relative to both rotor and stator blades and can excite destructive vibrations in either or both.

The reason for the propagation may be seen physically by considering the way the flow is diverted when one blade stalls. As sketched at the top of figure 5.23 the stalling blocks the flow, which must then divert to neighboring passages. The diversion *increases* the incidence on the blade in the direction of the tangential velocity and decreases it in the opposite one, so that the former tends to stall, carrying the stall in the direction of the tangential velocity. While a number of linearized analyses have predicted the speed of propagation of the stall cells fairly well, there is no satisfactory technique for predicting the number of stall cells; hence there is no technique for predicting the frequency.

As the weight flow is reduced below the design value, rotating stall usually begins as a "partial stall," that is, covering only a part of the blade span, and with a small number of cells. As weight flow is further reduced, the number of cells increases and the pressure ratio of the compressor begins to decrease, as in figure 5.23. At some weight flow an abrupt transition may occur, to full stall, in which the blades are stalled over their full span, and this leads to a further drop in pressure ratio. If the weight flow is now allowed to increase, the transition back to partial stall occurs at a higher weight flow than did the original transition, so that there is an hysteresis in the characteristic.

This hysteresis, coupled with the characteristics of the combustor or other system into which the compressor discharges, is one possible cause of *surge*. Suppose, for example, that the compressor discharges into a combustor cavity through a passage as shown in figure 5.24. The combination of volume and passage can act as a resonator (called a Helmholtz resonator); the volume plays the role of spring and the fluid in the passage that of mass, in a mechanical analogy. The oscillation induces flows to and fro in the passage, which would appear as increases and decreases in $W\sqrt{\theta}/\delta$ superimposed on some average value. If the compressor is operating at point 1 of figure 5.23 and the oscillator is set in motion by some perturbation so that a negative velocity (decrease of $W\sqrt{\theta}/\delta$) is imposed at the compressor outlet, the compressor will change suddenly to full stall, as indicated at zero time in figure 5.24. The lower pressure will persist until the oscillation leads to a positive perturbation in $W\sqrt{\theta}/\delta$, when the compressor changes to partial stall and delivers the higher pressure. Meanwhile, the pressure in the combustor p_4 fluctuates roughly $\pi/2$ out of phase with the velocity. Because p_3 is large when

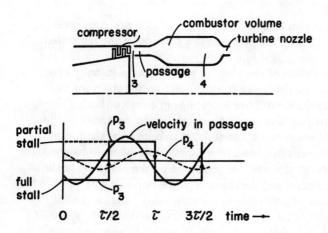

5.24 Mechanism of excitation of "surge" by the stall hysteresis shown in figure 5.23.

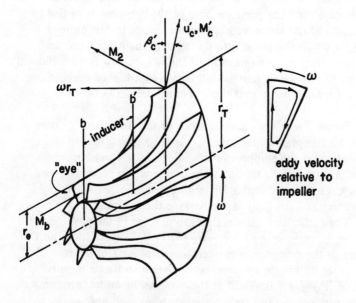

5.25 The rotor or impeller of a centrifugal compressor, showing the inlet and exit velocities, and at the right, the secondary flow generated in the impeller by the tendency for the flow to preserve zero absolute vorticity.

the fluid is moving into the volume and low when it moves out, the oscillation will grow in this simple picture. Actually, dissipation and other effects lead to an oscillation at fixed amplitude, called a "relaxation oscillation." The oscillation can build to such amplitudes that the combustor will blow out or, if prolonged, the compressor blading will fail; hence it must be avoided in routine engine operation.

Rotating stall is also very undesirable in normal operation because the reductions in weight flow and efficiency that it causes can result in severe over temperature in an engine. Also the hysteresis may make it difficult to recover from stall by reducing the throttle setting. An excellent discussion of these phenomena is given in reference 5.9.

5.7 Centrifugal Compressors

As its name implies, the centrifugal compressor achieves part of the compression process by causing the fluid to move outward in the centrifugal force field produced by the rotation of the impeller. This part of the pressure rise differs from the pressure rise in axial flow compressor rotors and stators; instead of arising from the exchange of kinetic energy for thermal energy in a diffusion process, it arises from the change in potential energy of the fluid in the centrifugal force field of the rotor. It is therefore less limited by the problems of boundary layer growth and separation in adverse pressure gradients. Probably for this reason the centrifugal compressor first attained a range of pressure ratio and efficiency useful for turbojet engines and was used in the Von Ohain engine (1939) and the Whittle engine (1941).

An impeller (rotor) for a centrifugal compressor is sketched in figure 5.25. The air enters through the "eye" near the axis, is turned to the radial direction, and brought to a tangential velocity near that of the rotor by the time it reaches the rotor tip. The essential feature is that *all the fluid* leaves the rotor at the *tip radius* rather than over a range of radii as in the axial flow compressor. If the flow Mach number is small in the impeller passages and the air has the tangential velocity of the rotor ωr at all radii as it flows outward, the pressure gradient in the radial direction is $dp/dr = \rho \omega^2 r$, and if the flow is isentropic so that $\rho/\rho_b = (p/p_b)^{1/\gamma}$, we find upon integrating

$$(p_c/p_b)^{(\gamma - 1)/\gamma} - 1 = (T_c/T_b) - 1 = \tfrac{1}{2}(\gamma - 1)M_T^2 \qquad (5.36)$$

where $M_T^2 = (\omega r_T)^2/\gamma R T_b$ is the square of the tip Mach number based on inlet temperature, as defined for the axial compressor. The impeller should produce this static pressure ratio even when there is very little flow through it. For $M_T = 1$ and $\gamma = 7/5$, for example, $T_c/T_b = 1.2$ and $p_c/p_b \approx 1.9$.

The air leaving the impeller has kinetic energy due to its tangential velocity as well as its small radial velocity. If this kinetic energy can be converted to thermal energy with small losses, a further pressure rise occurs just as in the stator of the axial compressor, but this process is subject to all the diffusion difficulties encountered in the axial compressor stator and more. Suppose first that the process is isentropic; then the pressure ratio would be

$$p_d/p_c = [1 + \tfrac{1}{2}(\gamma - 1)M_c^2]^{\gamma/(\gamma-1)}$$

where $M_c^2 = (\omega r_c)^2/\gamma R T_c = M_T^2(T_b/T_c)$. So

$$(T_d/T_c) - 1 = (p_d/p_c)^{(\gamma-1)/\gamma} - 1 = \tfrac{1}{2}(\gamma - 1)M_T^2/[1 + \tfrac{1}{2}(\gamma - 1)M_T^2], \qquad (5.37)$$

and the overall pressure ratio becomes

$$(p_d/p_b) = [1 + (\gamma - 1)M_T^2]^{\gamma/(\gamma-1)},$$

or the temperature ratio becomes

$$(T_d/T_b) - 1 = (\gamma - 1)M_T^2. \qquad (5.38)$$

Comparing (5.36) and (5.38), we see that half of the temperature rise of the stage occurs in the stator, and for high efficiency the static pressure ratio of the stator must equal that of the rotor. This requirement for matching the large pressure ratio of the rotor with an equally high pressure ratio in the stator limits the efficiency of the centrifugal compressor with radial impeller vanes.

The advantage of high pressure ratio is offset by the inherently low mass flow capacity per unit of frontal area that results from the radial flow geometry. Because the ratio of inlet flow area to frontal area varies as the square of the ratio of inlet tip radius to outlet radius, the mass flow capacity is a small fraction of that of an axial flow compressor of equal diameter; however, if the inlet is nearly equal in diameter to the tip, then the advantage of centrifugal compression is lost. Its low mass flow eliminated the use of centrifugal compressors in aircraft engines, except for some small turboprops, until recently. With the increase in cycle pressure ratios, it has found application as the high-pressure compressor in some engines with multiple shafts. The high pressure and high density of the air in the high-pressure compressors cause their flow area to be small compared to that of the inlet stages, so the diameter of the centrifugal stage is not limiting.

5.7.1 Stage Pressure Ratio and Diffuser Mach Number
The stagnation temperature ratio of the centrifugal stage is given most conveniently by the Euler equation (5.9). If there are no preswirl vanes, the inlet tangential velocity is zero, so $\beta_b = 0$. In general the fluid does not leave the impeller exactly radially, so $\tan \beta_c' \neq 0$. Even for radial vanes, there is some slip, that is, $\beta_c' > 0$. If we replace the axial velocity w_c by the radial velocity u_c' to obtain the tangential velocity relative to the impeller at its outlet, (5.9)

becomes

$$(T_{tc}/T_{tb}) - 1 = [(\omega r_c)^2/c_p T_{tb}][1 - (u'_c/\omega r_c) \tan \beta'_c]$$

or, in terms of Mach numbers,

$$\tau_c - 1 = \left\{ \frac{(\gamma - 1)M_T^2}{1 + \frac{1}{2}(\gamma - 1)M_b^2} \right\} \left\{ 1 - \frac{M'_c}{M_T}[1 + \frac{1}{2}(\gamma - 1)M_T^2]^{1/2} \tan \beta'_c \right\}. \tag{5.39}$$

It has been assumed that $M'_c = M'_b$, so that T_c/T_b is given by (5.36).

The Mach number at stator entrance is given by

$$M_c^2 = [(\omega r_c - u'_c \tan \beta'_c)^2 + (u'_c)^2]/\gamma R T_c,$$

and this can be written

$$M_c^2 = \frac{M_T^2}{1 + \frac{1}{2}(\gamma - 1)M_T^2} + (M'_c)^2 \sec^2 \beta'_c - \frac{2M_T M'_c \tan \beta'_c}{[1 + \frac{1}{2}(\gamma - 1)M_T^2]^{1/2}}. \tag{5.40}$$

This value of stator-inlet Mach number is shown, along with the ideal pressure ratio $\pi_c = \tau_c^{\gamma/(\gamma - 1)}$ from (5.39), in figure 5.26. For π_c above about 3, $M_c > 1$, but it does not exceed 1.5 even at a pressure ratio of 16, so it would

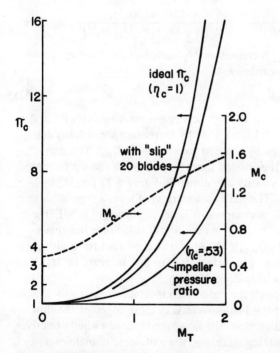

5.26 The ideal pressure ratio, stator-entrance Mach number, and impeller pressure ratio for a centrifugal compressor impeller as functions of tip Mach number.

seem possible to design efficient centrifugal compressors with pressure ratios of this order.

The impeller static pressure ratio given (5.36) is shown in figure 5.26 as a reasonable lower limit to π_c. If π_c is given by this relation, then $\eta_c = 0.53$ for $\beta'_c = 0$ and $M_b = 0.5$, as assumed in figure 5.26. This might be taken as a lower limit to the efficiency of the centrifugal compressor.

5.7.2 The Impeller

The function of the impeller is conventionally divided into two parts: bringing the air to the angular velocity of the rotor and carrying it radially outward while increasing its angular momentum and static pressure. The first function is performed by the inducer, the inlet portion of the rotor, of nearly constant tip diameter. It acts somewhat like an axial flow rotor without inlet guide vanes and with turning to the axial direction. If we apply the concept of diffusion factor to this part of the impeller, assuming that the flow velocity normal to the passage cross section is constant through the inducer (that the axial velocity is constant), the diffusion factor for the inducer can be written in terms of M_b, M_T, and the ratio of "eye" radius to tip radius,

$$D_{\text{ind}} = 1 - \frac{1}{[1 + (r_e/r_T)^2(M_T/M_b)^2]^{1/2}} + \frac{(M_T/M_b)(r_e/r_T)}{2\sigma[1 + (r_e/r_T)^2(M_T/M_b)^2]^{1/2}}. \quad (5.41)$$

This expression shows that D_{ind} increases as M_T increases for fixed r_e/r_T and vice versa. Since the mass flow capacity is

$$\dot{m}/\pi r_T^2(\rho u)_b^* \approx (A^*/A)(M_b)(r_e/r_T)^2 \quad (5.42)$$

based on tip frontal area, there is a conflict between mass flow capacity and pressure ratio. D_{ind} can be lowered somewhat by increasing the solidity but only to some limiting value given by the first two terms of D_{ind}. The variations of mass flow, divided by the choked mass flow through the tip frontal area, and the required inducer solidity are shown in figure 5.27 for $M_b = 0.5$, $M_T = 1.5$, and $D_{\text{ind}} = 0.5$. The mass flows are very low; a typical value of this parameter for an axial flow compressor is on the order of 0.5. Large values of σ_{ind} are required for r_e/r_T greater than about 0.4. Since the curve for σ_{ind} depends on the quantity $(r_e/r_T)(M_T/M_b)$, it is clear that for a given σ_{ind} the permissible r_e/r_T decreases proportionately to any increase in M_T/M_b so the mass flow decreases with increasing M_T.

As the air flows outward through the radial passages, its angular momentum is increased. Clearly, if the blades were very close together the flow would follow them, and for radial vanes the flow would leave radially relative to the rotor ($\beta'_c = 0$). As the spacing increases, the exit velocity inclines away from the direction of rotor motion ($\beta'_c > 0$) so that the work done by the impeller decreases. This is called "slip," and a slip factor is defined as the

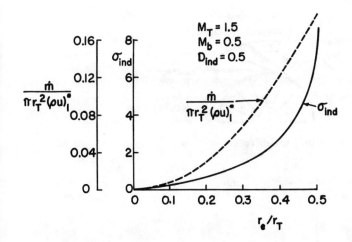

5.27   Mass flow capacity and inducer solidity required in a centrifugal impeller as functions of ratio of eye radius to tip radius.

ratio of actual tangential velocity to ωr_c. The slip has been estimated by calculations of incompressible flow through radial passages assuming that the flow remains irrotational in absolute coordinates as it passes through the rotor. For this to be so, it must have a vorticity opposite the angular velocity ω of the rotor, as sketched at the right in figure 5.25. The velocity backward relative to the tip leads to slip. Calculations by Stanitz (reference 5.10) gave slip factors of 0.90 for twenty blades and 0.93 for thirty blades. The effect of the slip of 0.90 on π_c is shown in figure 5.26.

5.7.3   The Diffuser

Several factors complicate the design of the diffuser. For pressure ratios above about 3 it must accept supersonic flow, with M_c as large as 1.4 at $\pi_c \approx 10$. To maximize the mass flow per unit of overall frontal area, the radial extent of the diffuser beyond the impeller tip should be as small as possible. But for this last requirement a "vaneless diffuser" could be used, in which the swirl velocity is decreased as the flow moves outward with constant angular momentum. A doubling of the radius would be required to halve the velocity in such a diffuser, and such a large area is usually unacceptable. As a compromise a short vaneless diffuser is often combined with a vaned two-dimensional diffuser, as sketched in figure 5.28a. Some recent experimental high pressure ratio compressors have used "pipe" diffusers formed by drilling axisymmetric channels nearly tangential to the rotor tip, as sketched in figure 5.28b (reference 5.11). These have given surprisingly good performance, better than that of equivalent two-dimensional diffusers. Although the

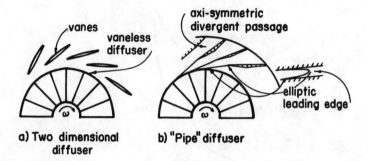

5.28   Two types of diffusers for centrifugal compressors. At the left, a short, vaneless diffuser followed by a two-dimensional vaned diffuser. At the right, a "pipe" diffuser.

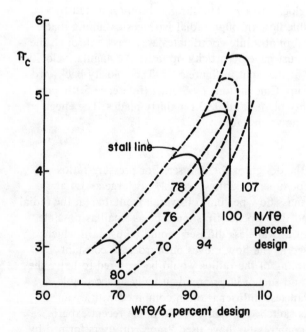

5.29   Performance map for a typical centrifugal compressor with design pressure ratio of 5.

reasons are not fully understood, they may be connected with the "sweep" of the contours presented to the supersonic flow. Sweep has been used for the leading edges of hypersonic ramjet diffusers.

These rapid advances make it difficult to estimate the efficiencies ultimately possible with centrifugal compressors at high pressure ratios. Currently, $\eta_c =$ 0.81–0.83 is attained at $\pi_c = 4$–7. It seems possible that π_c may be raised to 10–12 with η_c as large as 0.80–0.85 in a single stage.

A performance map for a typical centrifugal compressor with design pressure ratio of 5 is shown in figure 5.29.

References

5.1 F. E. Marble, "Three Dimensional Flow in Turbomachines," in *Aerodynamics of Turbines and Compressors* (High Speed Aerodynamics and Jet Propulsion Series, vol. 10), Princeton, 1964.

5.2 W. R. Hawthorne and R. A. Novak, "The Aerodynamics of Turbomachinery," *Annual Reviews of Fluid Mechanics*, Vol. 1 (1969), p. 341.

5.3 S. Lieblein, "Experimental Flow in Two-Dimensional Cascades," *Aerodynamic Design of Axial Flow Compressors*, NASA SP–36, 1965.

5.4 A. A. Mikolajczak, A. L. Morris, and B. V. Johnson, "Comparison of Performance of Supersonic Blading in Cascade and in Compressor Rotors," ASME Paper 70–GT–79.

5.5 J. E. McCune, *Journal of Aerospace Sciences*, Vol. 25, p. 544.

5.6 W. D. Rannie, "The Axial Compressor Stage," in *Aerodynamics of Turbines and Compressors* (High Speed Aerodynamics and Jet Propulsion Series, vol. 10), Princeton, 1964.

5.7 K. G. Harley and E. A. Burdsall, "High Loading Low-Speed Fan Study II: Data and Performance—Unslotted Blades and Vanes," NASA CR 72667 (PWA 3653).

5.8 J. P. Nikkanen and J. D. Brooky, "Single Stage Evaluation of Highly Loaded High Mach Number Compressor Stages V," NASA CR 120887 (PWA–4312).

5.9 E. M. Greitzer, "Surge and Rotating Stall in Axial Flow Compressors," *Engineering for Power*, Vol. 98, No. 2 (April 1976), p. 190.

5.10 J. D. Stanitz and G. O. Ellis, "Two Dimensional Compressible Flow in Centrifugal Compressors with Straight Blades," NACA TN 1932, 1949.

5.11 D. P. Kenny, "A Comparison of the Predicted and Measured Performance of High Pressure Ratio Centrifugal Compressor Diffusers," ASME Paper 72–GT–54.

Problems

5.1 Calculate the stagnation temperature ratio for a compressor rotor with $r_H/r_T \approx 1.0$, $M_T = 1.0$, $\sigma = 1$, $D = 0.5$, assuming the axial velocity is constant, no inlet guide vanes, and $M_b = 0.5$. If the polytropic efficiency is 0.9, find the stagnation pressure ratio. If the mass flow rate of this compressor rotor is 100 kg s^{-1}, what power is required to drive it?

5.2 When a gas turbine is used to power a helicopter, two alternatives are (a) to use a power turbine geared to the rotor (a turboshaft drive), or (b) to duct the engine exhaust gas through the rotor to nozzles at the blade tips, the jets driving the rotor. Compare these two schemes from the standpoint of efficiency of utilization of the energy of the gas generator exhaust gas:
1. Write the Euler equation to apply to the tip jet rotor drive.
2. Derive an expression for the specific fuel consumption for the tip jet drive, containing the rotor tip Mach number as a parameter.
3. Compare and plot the specific fuel consumption of the tip jet drive versus the geared turboshaft drive as a function of tip Mach number.

Assume all processes are ideal. For $P/\dot{m}c_pT_0$, you should find

$$(P/\dot{m}c_pT_0)_{\text{tip jet}} = (\gamma - 1)M_T^2[(1 + W^*/(\gamma - 1)M_T^2)^{1/2} - 1]$$

where P is the power delivered to the rotor by the tip jet drive, W^* is as defined in section 2.7 the value of $(P/\dot{m}c_pT_0)_{\text{shaft}}$ for the shaft drive system, and M_T is the rotor tip tangential Mach number. Note that this result implies $(P/\dot{m}c_pT_0)_{\text{tip jet}} \leq (P/\dot{m}c_pT_0)_{\text{shaft}}/2$. Why is this so?

5.3 A high-performance transonic compressor stage is to be designed to give an ideal stagnation pressure ratio of 1.50 at sea level static conditions with a tip tangential speed of 400 m s^{-1} and mass flow of 100 kg s^{-1}. It is to have a hub/tip radius ratio at the rotor inlet of 0.50, inlet guide vanes that produce a flow angle $\beta_a = 30°$ at the tip, and an axial Mach number = 0.5 at rotor inlet.

The result of your "design" should be (a) velocity triangles for the hub, tip, and mid radii; (b) sketches of the cascades at the same three radii, giving the flow angles, blade spacings/chord, and rough blade shapes; (c) a sketch of the annulus cross-section showing the contours of the hub and the casing and the blade locations.

You might proceed as follows:
1. From the Euler equation find the tangential velocity change across the rotor, hence β_c, and from the guide vane outlet angle, get A.
2. Assuming *incompressible* flow, find the axial velocity profiles at rotor inlet and exit by the method of 5.2.1 (note the utility of figure 5.5) and draw velocity triangles at hub, tip, and midspan.
3. Choose the solidity for rotor and stator so the largest "D factor" for each is about 0.5.
4. Now assuming that the axial Mach number is constant at 0.5, sketch the annulus cross-section.

5.4 A single-stage compressor produces a maximum pressure ratio of 1.5 with an efficiency $\eta_c = 0.85$ when operating in air at near ambient conditions (300°K and 1 atm). Its mass flow is 100 kg s^{-1}. If the rotative speed is held

constant, what maximum pressure ratio and mass flow could it be expected
to produce at inlet conditions of 1000°K and 0.5 atm? Would the efficiency
change?

5.5 A turbojet engine is to be optimized for vertical takeoff. The requirement
is that the engine operate for 5 minutes at full thrust. The *optimum engine*
minimizes the *sum of engine weight plus fuel weight* for the five-minutes'
operation.

From the information presented in figures 5.17 and 5.18, estimate the best
hub/tip radius ratio, tip tangential Mach number, and pressure ratio for the
compressor for this engine. Consider the following as parameters: π_c, n
(number of stages), M_T, r_H/r_T. Begin by writing expressions for specific
impulse and $F/\dot{m}a_0$ from the simplified cycle analysis, but including the effect
of η_c. Then estimate the compressor weight. Assume that the engine weight
is a constant times the compressor weight. Make any plausible assumptions
or estimates that seem necessary.

5.6 For a compressor with overall pressure ratio $\pi_c = 25$, estimate the number
of stages required as a function of tangential Mach number M_T in the first
stage, and plot this relationship. Use the results given in figure 5.18 for
$r_H/r_T = 1.0$. Assume that the tip diameter of all stages is the same, and that
the efficiency is unity.

5.7 Multistage, high-pressure ratio compressors often have variable-angle
stator blades in the first few stages to help relieve the low-speed operating
problems referred to in section 5.5. Sketch the velocity triangle for the inlet
stage of such a compressor with inlet guide vanes, and indicate which way
the inlet guide vanes and the first-stage stators should be rotated when the
compressor goes from full speed to idle.

5.8 With modern materials and methods of stress analysis, it is possible to
design centrifugal compressor impellers that have "backward curved" blades
($\beta_c' > 0$) and yet operate at large tip speeds, giving pressure ratios of the order
of 8. Efficiency can be improved by the reduction of M_c which results from
the backward curvature. To understand this trend, determine the variation
of M_c^2 with β_c' for fixed T_{tc}/T_{tb}. Is there a value of β_c that minimizes M_c^2?

6
Turbines

Much of what has been said about compressors applies equally well to turbines, but two factors lead to major differences between the turbine and compressor. The high gas temperature at turbine inlet introduces material problems much more serious than those of the compressor and has led to blade cooling in modern aircraft engines. The high temperature also leads to lower tangential Mach numbers for the turbine blades compared to compressor blades of the same radius, and this eases the aerodynamic problems somewhat. A second major difference is that the pressure falls through the turbine rather than increases as in the compressor. This dropping pressure thins the boundary layers, reducing separation problems and rendering the aerodynamic design less critical.

In turbojet engines the turbine efficiency is less critical to the engine performance than the compressor efficiency (see section 3.7), and this combined with other factors led to considerably less detailed aerodynamic development of the turbine than of the compressor. The situation has now changed because turbine efficiencies are critical in high bypass turbofans. In such engines the cost of the turbine can be a major part of engine cost (because of the difficult and expensive materials), and both its weight and its efficiency are important, so there is a great incentive to reduce the number of turbine stages while increasing efficiency.

To clarify the relationship between the turbine and compressor consider a turbojet engine with compressor and turbine diameters equal. The compressor-turbine power balance requires that $\theta_0(\tau_c - 1) = \theta_t(1 - \tau_t)$. As we saw, for a single compressor stage $\tau_s - 1 \propto M_T^2$; the constant of proportionality depends on blading geometry. The same is true for a multistage compressor, as examination of figure 5.22 shows, so we can write $\tau_c - 1 = M_{Tc}^2 f_c(\text{geom})n_c$, where M_{Tc} is the tangential Mach number of the compressor and n_c is the number of compressor stages. Similarly, $1 - \tau_t = M_{Tt}^2 f_t(\text{geom})n_t$, where M_{Tt} is the tangential Mach number of the turbine rotor. The power balance then becomes

$$\theta_0 M_{Tc}^2 f_c(\text{geom})n_c = \theta_t M_{Tt}^2 f_t(\text{geom})n_t.$$

For equal blade speeds $\theta_0 M_{Tc}^2 = \theta_t M_{Tt}^2$ so the effects of higher temperature and lower Mach number on the power capability of the turbine as compared to the compressor just cancel. If f_c were equal to f_t, the number of compressor and turbine stages would be equal. Actually, because of the better behavior of the boundary layers in the turbine, the blade loading can be higher than in the compressor, and $f_t > f_c$, with the result that $n_t < n_c$. In early axial flow turbojets the ratio of compressor to turbine stages was as high as 15. It is decreasing as the design of high-work compressor stages improves; a ratio of 4 or 5 is more typical of advanced engines.

Another requirement is that the mass flow of the turbine equal that of the compressor. This sets a minimum to the flow area at the turbine outlet (station 5 of figure 1.4) relative to the compressor inlet area A_2. If we assume that $M_5 = M_2$, the area ratio A_5/A_2 is simply $A_5/A_2 = (T_{t5}/T_{t2})^{1/2}(p_{t2}/p_{t5})$; for the ideal turbojet cycle this can be written in terms of π_c, θ_0, and θ_t as

$$A_5/A_2 = (\theta_t/\theta_0)^{1/2}/\pi_c [1 - (\{\pi_c^{(\gamma-1)/\gamma} - 1\}/(\theta_t/\theta_0))]^{(\gamma+1)/2(\gamma-1)}. \tag{6.1}$$

This result is plotted in figure 6.1. The turbine flow area limits frontal area only for very low π_c; but even at $\pi_c = 10$ it is about half the compressor flow area, and this has important implications for the turbine stress and stage pressure ratio.

Aircraft engines use axial flow turbines almost exclusively, as do most large stationary and automotive gas turbines. Radial flow stages, the analogue of the centrifugal compressor, are used in small turboshaft engines and in auxiliary power units. Only the axial flow turbine will be discussed here.

6.1   Turbine Stage Characteristics

A turbine stage consists of a nozzle row and a rotor. Sometimes a down-stream stator is added if there is only one stage. As indicated in figure 6.2, the nozzle vanes turn the flow while dropping the pressure and raising the Mach number. The rotor blades, moving in the same direction as the tangential velocity from the vanes, turn the flow back to remove the angular momentum put in by the vanes. They may simply turn the flow without a further drop in pressure, in which case the turbine is called an impulse turbine, or they may further drop the pressure. The ratio of the rotor pressure drop to the total stage pressure drop is called the *degree of reaction*. Velocity diagrams for impulse and 50 percent reaction stages are shown in figure 6.2.

The Euler equation (5.9) applies to the turbine stage as well as to the compressor stage. It is more conveniently written

$$1 - \frac{T_{tc}}{T_{tb}} = \left[\frac{(\omega r_b)^2}{c_p T_{tb}} \right] \left[\frac{w_b}{\omega r_b} \tan \beta_b + \left(\frac{r_c}{r_b} \right)^2 \left(\frac{w_c}{\omega r_c} \tan \beta_c' - 1 \right) \right], \tag{6.2}$$

so that both sides are positive, and the last term in the square brackets represents v_c, the tangential velocity at the rotor outlet. Usually this is small at the design condition because any swirl energy left in the exhaust gas as it leaves the engine nozzle is lost. Thus for the sake of simplicity we will assume $\tan \beta_c' = \omega r_c/w_c$ and write

$$1 - \tau_t = \{[(\gamma - 1)M_T M_b]/[1 + \tfrac{1}{2}(\gamma - 1)M_b^2]\} \sin \beta_b, \tag{6.3}$$

where $M_b = V_b/(\gamma R T_b)^{1/2}$ is the total Mach number leaving the nozzle vanes.

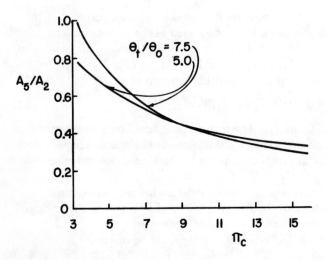

6.1   Ratio of turbine outlet area to compressor inlet area for a turbojet engine (or gas generator) with equal compressor and turbine diameters, as a function of compressor pressure ratio.

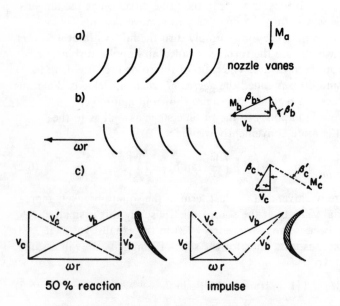

6.2   Construction of velocity triangles for a turbine stage, with typical composite diagrams for a 50% reaction stage and an impulse stage.

This form of the Euler equation shows that for maximum work per stage (small τ_t) we want large M_t (high blade speed), large M_b, and large turning in the vanes (large $\sin \beta_b$).

6.1.1 Degree of Reaction, Stage Pressure Ratio, Blade Temperature, and Mass Flow

We will define a degree of reaction R as the ratio of kinetic energy change in the rotor, relative to the rotor, to the sum of that change and the change in the vanes. That is

$$R = [(V_c')^2 - (V_b')^2]/[(V_c')^2 - (V_b')^2 + V_b^2 - V_a^2].$$

Assuming that the axial velocity is everywhere constant and equal to w_0 we get

$$R = [\sec^2 \beta_c' - \sec^2 \beta_b']/[\sec^2 \beta_c' - \sec^2 \beta_b' + \sec^2 \beta_b - 1].$$

Using the condition $\tan \beta_c' = \omega r_c/w$, we can simplify this to

$$R = 1 - [M_b \sin \beta_b/2M_T] \tag{6.4}$$

so that (6.3) can be written in terms of R as

$$1 - \tau_t = \{[(\gamma - 1)M_T^2]/[1 + \tfrac{1}{2}(\gamma - 1)M_b^2]\}2(1 - R). \tag{6.5}$$

For a given blade speed, the *impulse* turbine ($R = 0$) gives the *largest temperature drop*. On the other hand, because there is little pressure drop in the rotor, viscous effects tend to be worse for low degrees of reaction, so the efficiency is lower for low reaction than for about 50 percent reaction ($R = 0.5$).

Another factor controlled by the degree of reaction is the difference between the turbine inlet stagnation temperature (which we have characterized by θ_t) and the stagnation temperature felt by the rotor blades. This temperature is relevant to the turbine rotor stress problem. Denoting the stagnation temperature in *rotor* coordinates T_{tr}', it is

$$T_{tr}' = T_b[1 + \tfrac{1}{2}(\gamma - 1)(M_b')^2] = T_{tb}[1 + \tfrac{1}{2}(\gamma - 1)(M_b')^2]/[1 + \tfrac{1}{2}(\gamma - 1)M_b^2],$$

and since $T_{tb} = T_{ta}$,

$$\frac{T_{tr}'}{T_{ta}} = \frac{1 + \tfrac{1}{2}(\gamma - 1)M_b^2 \cos^2 \beta_b \sec^2 \beta_b'}{1 + \tfrac{1}{2}(\gamma - 1)M_b^2} = 1 + \frac{\tfrac{1}{2}(\gamma - 1)M_T^2}{1 + \tfrac{1}{2}(\gamma - 1)M_b^2}(4R - 3). \tag{6.6}$$

The second equality results from a little manipulation using (6.4); it shows that the rotor temperature increases with increasing R. Taking $M_T = 0.5$, $M_b = 1$ shows that $T_{tr}'/T_{ta} = 1 + 0.042(4R - 3)$, and the difference between an impulse turbine ($R = 0$) and a 50 percent reaction turbine ($R = 0.5$) is 0.08 or about $100°$K at $T_{ta} = 1400°$K. In terms of the changes possible with improved materials, this is a very large difference; it is equivalent to about ten years' effort on alloys.

The mass flow per unit of annulus area is also related to the degree of reaction for a given blade speed. To see this we must examine the compressible flow turning in the nozzles. Suppose that the vanes have zero thickness and that the radial height of the passages is constant, so that the flow is two dimensional in first approximation; we can deduce a connection between β_b and M_b. Continuity of the axial flow requires that $\rho_b V_b \cos \beta_b = \rho_a V_a$ while the constancy of T_t gives $T_b[1 + \frac{1}{2}(\gamma - 1)M_b^2] = T_a[1 + \frac{1}{2}(\gamma - 1)M_a^2]$ and for isentropic flow $(\rho_b/\rho_a) = (T_b/T_a)^{1/(\gamma - 1)}$. Combining these we find

$$\cos \beta_b = (M_a/M_b)\{[1 + \frac{1}{2}(\gamma - 1)M_b^2]/[1 + \frac{1}{2}(\gamma - 1)M_a^2]\}^{(\gamma + 1)/2(\gamma - 1)}.$$

For a given M_a, β_b first increases ($\cos \beta_b$ decreases) as M_b is increased; then β_b decreases again for M_b large (because $(\gamma + 1)/(\gamma - 1) > 1$) so there is a value of M_b for which β_b is largest and a value for which $M_b \sin \beta_b$ is largest for a given M_a. Since the mass flow capacity increases as M_a increases (up to 1), a compromise must be made between stage temperature ratio, which improves with increasing $M_b \sin \beta_b$, and mass flow capacity. From the expression for $\cos \beta_b$

$$M_b^2 \sin^2 \beta_b = M_b^2 - M_a^2\{[1 + \frac{1}{2}(\gamma - 1)M_b^2]/[1 + (\gamma - 1)M_a^2]\}^{(\gamma + 1)/(\gamma - 1)}.$$

Differentiating with respect to M_b and putting the result to zero, we find the M_b that maximizes $M_b \sin \beta_b$ for fixed M_a, and substituting this into $M_b^2 \sin^2 \beta_b$ gives

$$
\begin{aligned}
(M_b^2 \sin^2 \beta_b)_{max} &= \frac{2}{\gamma - 1}\left[[1 + \frac{1}{2}(\gamma - 1)M_a^2]\left(\frac{1 + \frac{1}{2}(\gamma - 1)M_a^2}{\frac{1}{2}(\gamma + 1)M_a^2}\right)^{(\gamma - 1)/2} - 1\right] \\
&\quad - M_a^2\left[\frac{1 + \frac{1}{2}(\gamma - 1)M_a^2}{\frac{1}{2}(\gamma + 1)M_a^2}\right]^{(\gamma + 1)/2} \\
&= \frac{2}{\gamma + 1}(M_b^2 - 1) \\
&= (1 - \tau_t)^2\left[\frac{1 + \frac{1}{2}(\gamma - 1)M_b^2}{(\gamma - 1)M_T}\right]^2 = [2M_T(1 - R)]^2.
\end{aligned}
$$

(6.7)

The last equalities follow from (6.3) and (6.5). This relation is plotted in figure 6.3, along with the mass flow density, which is just $A^*/A(M_a)$. We see that as M_a increases, $M_b \sin \beta_b$ decreases, with a corresponding decrease in $1 - \tau_t$, but the mass flow density, of course, increases. For a given M_T the impulse turbine has lower M_a than the 50 percent reaction turbine, hence lower mass flow to offset its greater work.

M_b is always greater than unity, as indicated by the second equality of (6.7)

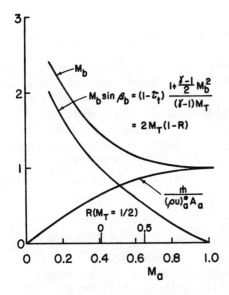

6.3 Maximum tangential Mach number and mass flow as functions of inlet axial Mach numbers (and of reaction for fixed blade speed).

and plotted in figure 6.3. The optimum nozzle therefore always is convergent-divergent; within the given assumptions it turns to the value of β giving $M = 1$, then back to a lower value. For $M_a > 0.5$, however, the optimum M_b is not much above 1, so a simple convergent nozzle is near optimum.

The optimum degree of reaction for any particular application will depend on the relative importance of efficiency, temperature drop per stage, mass flow capacity, and blade temperature. Degrees of reaction near 50 percent are usually preferable in subsonic cruise engines where efficiency is a dominant requirement, but where engine weight must be minimized the impulse turbine may be better.

6.1.2 Radial Variations

In fact, the above comments are oversimplified because the degree of reaction must vary with radius if the exit swirl velocity is to be small at all radii and the turbine is to produce a uniform temperature drop over its full annulus. If we put $M_T = M_T(r_T)\zeta$ where ζ is r/r_T and then interpret $M_T(r_T)$ as the *tip* tangential Mach number, the reaction must be $R = 1 - (1 - R_T)/\zeta^2$, from (6.5), where R_T is the reaction at the tip. If, for example, $R_T = 0.5$ and $r_H/r_T = 0.707$, the hub will have $R = 0$, corresponding to an impulse design.

Because this rapid variation of reaction with radius may lead to unacceptable efficiencies at the hub for low values of r_H/r_T, it is sometimes necessary to accept exit swirl from the rotor, either providing exit vanes to remove it or allowing it to carry through the next turbine stage just as the inlet guide vanes wirl did in the compressor.

When the flow differs from a free vortex, the angular velocity changes across blade rows induce radial variations of axial velocity just as in the compressor. The particular case of constant $(1 - \tau_t)$ and zero exit swirl corresponds to a free vortex, so these effects are small for such a design, but designs with nonzero swirl leaving the rotor will in general lead to streamline shifts. Of particular practical importance is the case for β_b equal to a constant because this allows the nozzle vanes to be of constant shape over their length.

6.2 Turbine Blading

The requirements imposed on turbine blading by cooling, such as larger than otherwise desirable leading and trailing edge thicknesses, and the problems of manufacture from refractory alloys tend to dictate against great aero-dynamic refinement in turbine blading. The available systematic data for turbine cascades are for uncooled configurations; thus these data are not directly applicable to the cooled stages of a turbine. They are applicable to the un-cooled stages, however, and in turboshaft engines or high-bypass turbofan engines the efficiency of these stages is central to engine performance.

A comprehensive survey of the many attempts to develop correlations of turbine efficiency can be found in reference 6.1. Three main sources of loss are usually identified exclusive of those associated with cooling: profile losses, due to viscous shear on the blade sections; secondary flow losses, similar to those suffered in compressors, and discussed in section 5.2.5; and losses due to leakage past the blade tips.

The leakage is more critical in turbines than in compressors because of the generally larger pressure difference across a blade row. Many turbine rotors are shrouded, that is, the blade tips carry an annular ring made up of segments, one on each blade, and with a knife-edge seal that runs close to an abradable material such as honeycomb. Such seals are used in the engines shown in figures 1.15 to 1.19. The efficiency is decreased 1 percent for each percent of leakage past the rotor, but unfortunately there are no simple means available for calculating the leakage. It is estimated by taking the blade clearance gap as an orifice in parallel with the rotor.

Methods for estimating secondary flow losses are the subject of active research but none are sufficiently developed to be useful in estimating turbine efficiency. Both tip clearance and secondary flow losses are included in correlations of efficiency as a function of aspect ratio (the ratio of blade span to blade chord). Such correlations are reviewed in reference 6.1.

The profile losses in a turbine are due principally to two effects, the adverse pressure gradient on the suction side and shock-boundary layer losses due to locally supersonic flows on the suction side. Losses due to the first of these have been correlated in a variety of ways, versus flow deflection angles, and solidities. Again, a comprehensive review is given in reference 6.1. A correlation versus the diffusion factor suggested in reference 6.2 is appealing in view of its successful application to compressors. For the turbine

$$D = 1 - (V_2/V_1) + |(v_2 - v_1)/2\sigma V_1|,$$

where, just as for the compressor the term V_2/V_1 represents the effect of mean velocity change through the blade passage, here $(v_2 - v_1)/2\sigma V_1$ represents the adverse pressure gradient on the suction side due to the blade force required to produce the flow deflection. Generally for a turbine $V_2/V_1 \geqslant 1$; it is unity for impulse blading and larger for 50 percent reaction (see figure 6.2). Thus the flow acceleration reduces D while the deflection increases it. The turbine cascade loss data of reference 6.3 is presented in the format of $(\bar{\omega}_2 \cos \beta_2/2\sigma)$ $(\cos \beta_2/\cos \beta_1)^2$ versus D in figure 6.4. Here we choose $\bar{\omega}_2 \equiv (\bar{p}_{t2} - p_{t1})/(p_{t2} - p_2)$ as the loss parameter rather than $\bar{\omega}_1 \equiv (p_{t2} - p_{t1})/(p_{t1} - p_1)$, so that the stagnation pressure loss is measured against the largest dynamic pressure. The figure includes impulse blading and 50 percent reaction blading

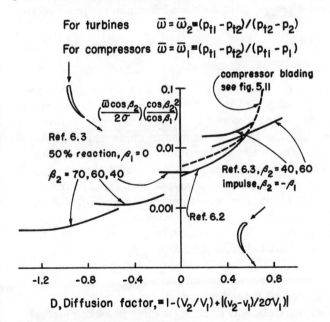

6.4 Turbine cascade loss data presented as "D" factor correlation, and compared to compressor cascade data.

with a variety of deflection angles and solidities $1 < \sigma < 3$. The compressor loss correlation is included for comparison as is the turbine correlation of reference 6.2. Impulse blading has D values in the general range of compressor blading and correspondingly large loss factors. Reaction blading has negative D for the range of solidities covered in the data and correspondingly small loss factors corrected for the flow angles.

The corrected loss factor plotted in figure 6.4 measures the ratio of boundary layer momentum thickness θ^* to chord. This does become very small for negative D as our physical arguments suggest, but this *does not* mean that $\bar{\omega}$, which measures the fractional stagnation pressure drop in the cascade, is small, because the negative D is attained by turning to large β_2. For $\beta_2 = 70°$, $\cos^3 \beta_2 \approx 0.04$, so that $\bar{\omega}_2 = 50\sigma$ times the value plotted in figure 6.4. For $\sigma = 1$, which corresponds to $D = -0.06$, $\bar{\omega}_2 = 0.05$.

Even though blade inlet Mach numbers are generally small, shock losses arise from locally high Mach numbers in the turbine blades. As the blade loading is increased, the velocity and Mach number over the suction surface rise, eventually reaching supersonic values. When this happens there is an embedded region of supersonic flow, which can shock, as sketched in figure 6.5. The shock can in turn cause boundary layer separation in the adverse pressure gradient on the aft portion of the blade; the net result is a large increase in loss. This factor sets the limit on blade loading in modern aircraft turbines.

6.3 Turbine Cooling

The desire for higher thrust per unit of air flow provides a powerful incentive to increase the turbine inlet temperature. Fuel consumption also improves if the increased temperature is accompanied by an increase in compressor pressure ratio. Over the years there has been a gradual improvement in materials, permitting small increases in temperature in new engines and uprating of existing ones. This trend is shown in figure 6.6. About 1960 the introduction of air-cooled turbines gave a small initial increase in T_{t4} and increased the rate of improvement with time. The latest commercial transport engines (JT9-D, CF-6, RB-211) have turbine inlet temperatures *at takeoff power* near $1550°K$. The turbine inlet temperatures at cruise are somewhat lower.

Some experimental engines are now operating with near-stoichiometric temperatures (in the range of $2000°K$ to $2500°K$), so that there is a considerable margin left for improvement in service engines.

Of the many schemes proposed for turbine cooling (including liquid thermosiphon and boiling systems), only direct air cooling has seen practical application. Air is bled from the compressor, carried aft, and introduced into the turbine (rotor) blades through their roots, as sketched in figure 6.7.

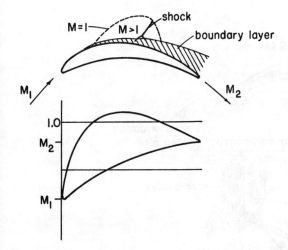

6.5   Mach number distribution and shock-boundary layer interaction on highly loaded turbine blade.

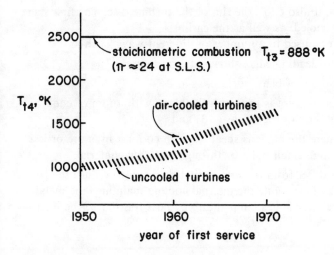

6.6   Trend of turbine inlet temperature with time.

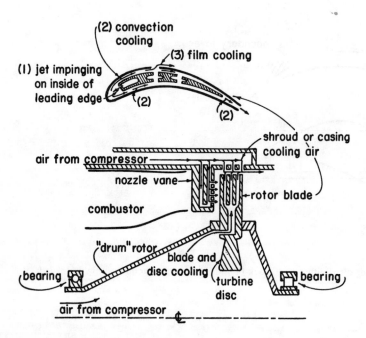

6.7   Schematic of air-cooled turbine, with cross-section of cooled airfoil section at top.

In entering the blades, it also cools the rim of the turbine disc. The first stage nozzle vanes are also cooled, as well as the casing.

The cooling air can be used in a number of different ways, such as impingement cooling of leading edge, shown at 1 on figure 6.7;

convective internal cooling of blade surface, as at 2;

film cooling, where the attempt is to sheath the blade with a film of cool air introduced through holes or a slit, as at 3; and

aspiration cooling, where the blade is sheathed with cool air by more-or-less uniform injection through small holes or through a porous material.

The last two methods differ from the first in that they aim to reduce the heat transfer to the blade surface, while the internal cooling maintains the metal temperature below the outside gas temperature by heat transfer to the internal cooling air.

6.3.1   Internal Cooling

Consider first the internal cooling scheme. The rate of heat transfer from the gas to the blade surface is expressed in terms of a Stanton number St defined so that if q_w is the rate of heat transfer to the surface, then

$$q_w = St[\rho u c_p (T_r - T_w)] \tag{6.8}$$

where T_r is the "adiabatic recovery temperature" (the temperature attained by an insulated surface in the flow) and T_w is the wall temperature. For our purposes, we may take T_r to be the stagnation temperature *relative to the blade* (see section 6.1). The heat transferred to the blade surface must be conducted through the metal, so that if k_s is the thermal conductivity, $q_w = -k_s$ grad T where T is the temperature in the metal. The Stanton number has a particularly useful physical interpretation. The total heat transferred to a blade of chord c per unit span is of order $cq_w = c[\rho u c_p(T_r - T_w)]St = \sigma[\rho u c_p(T_r - T_w)s]St$, and the quantity in brackets is the total thermal energy flux through a blade channel measured relative to the blade temperature. Thus we can say that the Stanton number (times solidity) gives the ratio of the heat transferred to the blade to the heat that would result from cooling all flow through the blade passage to blade temperature.

The Stanton number is controlled by the boundary layer behavior. If the Prandtl number is unity, or if in turbulent flows the Reynolds analogy applies, then $St \approx C_f/2$ where C_f is the friction coefficient. For flow over a flat plate

$$2St = C_f = 0.664(Re_x)^{-1/2} \quad \text{(laminar)},$$

$$2St = C_f = 0.0592(Re_x)^{-1/5} \quad \text{(turbulent)}. \tag{6.9}$$

Although these relations do not apply quantitatively to the turbine blade, they indicate trends. A detailed discussion of turbine blade heat transfer will be found in reference 6.4.

We see that if the boundary layer is laminar $q_w \propto x^{-1/2}$, while if it is turbulent $q_w \propto x^{-1/5}$. Of course, q_w is not infinite at the leading edge of the blade, but it can be very large, hence the impingement cooling. For $p_{t4} = 20$ atm and $\theta_t = 6$, $Re_x = 3 \times 10^7$ per meter at $M = 1$. (A useful fact to remember is that $Re_x = 1.2 \times 10^7$ per meter for air with $T_t = 288°K$ and $P_t = 1$ atm at $M = 1$.) Transition from laminar to turbulent flow occurs for Re_x between 3×10^5 and 10^6 on a flat plate. The Reynolds number for transition on a turbine blade is difficult to predict but probably lies below 10^6, so that transition will occur on most turbine blades. When it does the heat transfer rate increases, so that the distribution of q_w on a blade is somewhat as sketched in figure 6.8.

The problem of designing the cooling system is to schedule the internal flow so that with such a distribution of q_w the blade has as nearly a uniform temperature as possible. Because the thermal stresses caused by nonuniform temperature can be a limiting factor, it is important to properly schedule the cooling air. Indeed, if we set a limit on the *temperature difference* between

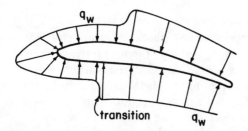

6.8 Distribution of heat transfer on a cooled turbine blade, showing sudden increase at boundary-layer transition to turbulent state.

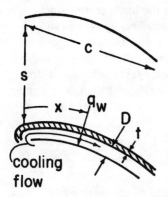

6.9 Schematic of internal cooling of a turbine blade.

two parts of the airfoil, then the precision with which the cooling can be controlled determines the permissible difference between gas temperature and blade temperature, hence the turbine inlet temperature, because the variations in blade temperature are a fraction of the gas-blade temperature difference.

The heat transfer to the cooling fluid, at points such as 2 in figure 6.7 can be estimated from results for flow in long tubes, such as

$$2St = C_f = 0.023 Re_D^{-1/5} \tag{6.10}$$

where the Reynolds number is now based on the hydraulic diameter of the passage (one-fourth the area divided by the circumference).

To see how the various requirements on the cooling system interact, consider the schematic cooled blade shown in figure 6.9 where the blade surface is cooled by flow through a passage of height D and the blade has a chord c and spacing s in a cascade. Suppose the cooling air enters the blade at temperature T_c. The total heat transferred to the blade (per unit span) is estimated as $cq_w = \sigma St[\rho u c_p(T_r - T_w)s]$; this must be transferred to the cooling air, with mass flow $\dot{m}_c$ (per unit span), so $cq_w = \dot{m}_c c_p(T_w - T_c)$ if the cooling air is heated all the way to blade temperature. It follows that the ratio of cooling to turbine mass flow must be at least

$$(\dot{m}_c/\rho u s) \geq \sigma St[(T_r - T_w)/(T_w - T_c)].$$ (6.11)

A typical value of St is 0.005 (C_f is about equal to θ^*/c of figure 5.9). Now for $T_r - T_w = 200\,°K$ and $T_w - T_c = 400\,°K$, we find $\dot{m}_c/\rho u s = 0.0025$; but because there are two sides to the blade, the required cooling flow according to this estimate would be about 0.005 of the turbine mass flow. The actual cooling flow is somewhat larger than this, partly because the cooling air must still be colder than the blade when it leaves it and partly because of the difficulty of getting sufficient heat transfer surface in the cooling passages. A value of about 3 times that given by (6.11) is representative of the state of the art in 1976.

6.3.2 Film and Transpiration Cooling

The objective in film and transpiration cooling is to cover the surface of the blade with a film of cool air, blocking the heat transfer to the surface. Ideally, such cooling eliminates the constraints imposed by heat transfer through the blade material and the resulting thermal stresses. In practice the film cooling produces a thermal stress pattern of its own. Introduction of air through the surface of the blade also disrupts the boundary layer flow, tending to increase both the shear (pressure loss) and heat transfer. The overall gain results from a balance of these undesirable effects against the heat transfer reduction.

In film cooling the cool air is introduced through a series of small holes inclined at an angle to the surface, as shown in figure 6.7. Their spacing is usually several hole diameters, the diameter being 0.05 to 0.1 cm or larger. Since the thickness of the boundary layer is generally of this order or less, the flow picture is one of discrete jets that penetrate the boundary layer, are turned by the free stream, and then flow along the surface, mixing with the boundary layer fluid. In transpiration cooling the scale of the holes and their spacing would be small compared to the boundary layer thickness so that in this case the cool air would be introduced into the low-velocity portion of the boundary layer. But such a fine distribution of pores is difficult to achieve and would be highly susceptible to plugging and surface damage. No such ideal transpiration cooling has been achieved in practice.

The two conflicting effects of boundary-layer disruption and cool air are shown clearly in figure 6.10, where the heat flux to a film-cooled surface is plotted as a function of $m = \rho_c u_c / \rho_\infty u_\infty$, where $\rho_c u_c$ is the mass flow density from the coolant holes and $\rho_\infty u_\infty$ is that in the external flow. Three curves are shown. In all cases the wall was at 555°K and the external flow at 1666°K. For the upper curve the "cooling" air was at stream temperature.

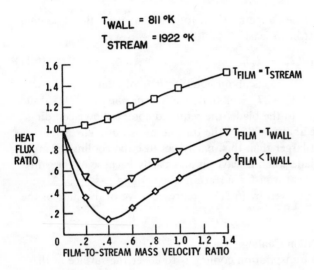

6.10   Effects of film-cooling mass flow and cooling airflow temperature on heat flux to film-cooled flat plate (from reference 6.5).

We see that increasing the cooling flow increases the heat flux, partly due to heat transfer in the cooling holes and partly due to stirring of the boundary layer. When the cooling air is at wall temperature, heat flux is reduced for small cooling flows but increased when the stirring effect becomes dominant at high flows. For a cooling air temperature below wall temperature the trends are similar, but the cooling is more effective.

The effectiveness of film cooling can be described quantitatively in a number of ways; the most common is in terms of an "adiabatic film effectiveness," defined as

$$\eta_{ad} \equiv (T_r - T_{rf})/(T_r - T_c) \tag{6.12}$$

where T_r is the adiabatic wall temperature (the temperature the wall would reach if insulated from all but the free stream) and T_{rf} is the temperature the wall will reach with the film cooling but no other cooling. As before T_c is the cooling air temperature. We see that η_{ad} varies between 0 (no cooling by film) and 1 (wall cooled to T_c). A heat transfer film coefficient is then defined as

$$h \equiv q_w/(T_{rf} - T_w). \tag{6.13}$$

The designer wishes to find q_w, having T_r, T_c, the mass flow ratio $m = \rho_c u_c/\rho_\infty u_\infty$, and the geometry of the cooling holes. From a series of experiments a correlation of η_{ad} such as that given in figure 6.11 would be found; the parameters are distance downstream from the injection holes and $m =$

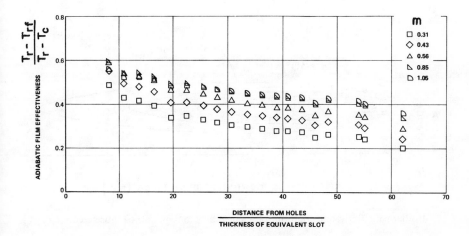

6.11   Adiabatic film effectiveness as a function of blowing parameter $m = \rho_c u_c / \rho_\infty u_\infty$ and distance downstream from holes divided by thickness of slot with same flow area as row of holes. For subsonic nozzle blade, two rows of holes .484 and .584 inches from leading edge on suction surface, total length of suction surface 2.2 inches (from reference 6.4).

$\rho_c u_c / \rho_\infty u_\infty$. But to get q_w requires some knowledge of h, which is usually taken to be that for a well-developed boundary layer with no injection. The mixing effects of the injection make this step uncertain.

An alternative approach is to define an "isothermal effectiveness" η_{iso} directly in terms of the heat transfer rates,

$$\eta_{iso} \equiv (q_{w0} - q_w)/q_{w0} \tag{6.14}$$

where q_w is the heat transfer rate with film cooling and q_{w0} is that for the same wall temperature without film cooling. Correlation of η_{iso} in terms of $m = \rho_c u_c / \rho_\infty u_\infty$, distance from the holes, Reynolds number, and perhaps other parameters should then include all effects of the injection. The "effectiveness footprint" of a single hole is shown in this form in figure 6.12 for a free-stream Mach number $M_\infty = 0.5$, an injection angle of 30° to the surface, and $m = 0.4$. The reasonably effective cooling is seen to extend only about ten hole-diameters downstream, even on the hole centerline. The effectiveness is shown as a function of m for a single hole in figure 6.13 and also for a double staggered row of holes as indicated on the figure. The "best" m is equal to about 0.4 as in figure 6.10.

Although attempts have been made to correlate η_{ad} (and η_{iso}) over wide ranges of temperature ratio, Reynolds number, and so forth, there seems to be no general correlation of sufficient accuracy for design purposes. Thus the design procedure must rely largely on experience with particular configurations combined with data such as that in figures 6.10, 6.11, or 6.13.

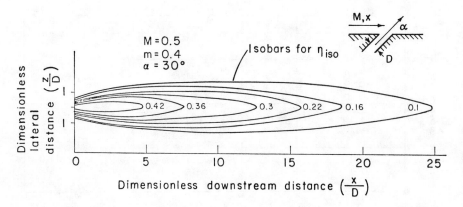

6.12 Isothermal effectiveness footprint of a single film cooling hole (from reference 6.6).

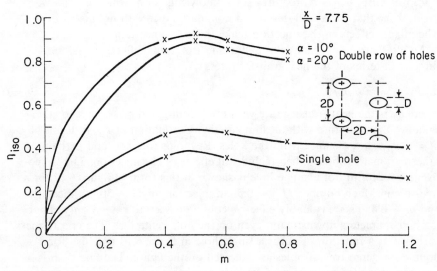

6.13 Isothermal effectiveness as a function of mass flow ratio at a distance of $x/d = 7.75$ for a single hole, and for a double row of holes (from reference 6.6).

6.3.3   The Effect of Cooling on Turbine Efficiency

To deal with the cooled turbine the previous definition of efficiency must be modified somewhat. We shall redefine it as the ratio of the *actual* turbine work per unit of *total* airflow, primary plus cooling, divided by the *ideal* work that would be attained in expanding that total airflow through the actual pressure ratio. We assume the cooling flow expands through the same pressure ratio as the primary flow. Thus if ε is the ratio of cooling airflow to total airflow,

$$\eta_t = \frac{(1 - \varepsilon)(T_{t4} - T_{t5}) + \varepsilon(T_{tc} - T_{t5})}{[(1 - \varepsilon)T_{t4} + \varepsilon T_{tc}][1 - (p_{t5}/p_{t4})^{(\gamma - 1)/\gamma}]} \tag{6.15}$$

where T_{tc} is the stagnation temperature of the cooling air and T_{t5} is the *mean* outlet temperature for the total flow. Given the temperatures and ε, π_t can be computed from this relation if η_t is known.

The cooling flow can influence η_t in three ways:

1. The cooling air emerging from the blades can change their drag characteristics, probably increasing drag.

2. The cooling air itself suffers a pressure loss in passing through the cooling passages, so that it has lower stagnation pressure when mixed into the downstream flow, thus decreasing π_t for a given τ_t.

3. The entropy of the flow as a whole is increased by the transfer of heat from the hot primary flow to the cooling flow.

Consider the third mechanism. The increase in entropy per unit mass of total flow that results from the transfer of an amount of heat ΔQ from the primary flow at turbine inlet temperature T_{t4} to cooling air temperature T_{tc} is $\Delta s = \Delta Q[(1/T_{tc}) - (1/T_{t4})]$. Taking the estimate of section 6.3.1 for the heat transfer gives

$$\Delta s_c/c_p = 2\sigma St[(T_{t4}/T_{tc}) - 1][1 - (T_w/T_{t4})] \tag{6.16}$$

per blade row; that is, we must double this value for a cooled stage.

As explained in section 5.2.4, the entropy change in the stage is related to the stagnation pressure and temperature changes across it by

$$\Delta s = (1 - \varepsilon)c_p \log(T_{t5}/T_{t4}) + \varepsilon \log(T_{t5}/T_{tc}) - R \log(p_{t5}/p_{t4})$$

so that

$$(p_{t5}/p_{t4}) = \{e^{-(\Delta s/c_p)}[(T_{t5}/T_{t4})^{1-\varepsilon}(T_{t5}/T_{tc})^{\varepsilon}]\}^{\gamma/(\gamma - 1)} \tag{6.17}$$

expresses the effect of the heat transfer on the turbine pressure ratio.

To account for the effect of pressure drop in the cooling air (the second mechanism), we may add the entropy change due to this pressure drop to the change due to cooling. This entropy change can be written

$\Delta s_f/c_p = -[(\gamma - 1)/\gamma]\varepsilon \log[(p_{t4} - \Delta p_f)/p_{t4}] = -\log[1 - (\Delta p_f/p_{t4})]^{\varepsilon(\gamma-1)/\gamma}$,

and since $\varepsilon \ll 1$,

$\Delta s_f/c_p \approx [(\gamma - 1)/\gamma]\varepsilon(\Delta p_f/p_{t4})$.   (6.18)

Now substituting (6.17) into (6.15), we find

$$\eta_t = \frac{1 - \tau_t - \varepsilon[1 - (T_{tc}/T_{t4})]}{1 - \tau_t \exp(-\Delta s/c_p) - \varepsilon[1 - (T_{tc}/T_{t4})]};$$

and with $\Delta s/c_p = (\Delta s_c + \Delta s_f)/c_p \ll 1$,

$$\eta_t = 1 - \left(\frac{\tau_t}{1 - \tau_t}\right)\left(\frac{\Delta s}{c_p}\right)$$

$$= 1 - \left[\frac{\tau_t}{1 - \tau_t}\right]\left[2\sigma St\left(\frac{T_{t4}}{T_{tc}} - 1\right)\left(1 - \frac{T_w}{T_{t4}}\right) + \left(\frac{\gamma - 1}{\gamma}\right)\varepsilon\left(\frac{\Delta p_f}{p_{t4}}\right)\right].$$

(6.19)

The significance of this relation is best seen by means of a typical example. Take $\sigma = 1$, $St = 0.005$, $T_w/T_{t4} = 0.7$, $T_{t4}/T_{tc} = 2.5$, $\tau_t = 0.82$ ($\pi_t = 0.5$), and $\Delta p_f/p_{t4} = 0.5$. From (6.11) we estimate $\varepsilon \approx 0.01$, and η_t becomes $\eta_t \approx 1 - (0.82/0.18)[(0.01)(1.5)(0.3) + (0.01)(0.14)] = 1 - 0.027$; thus there is a 2.7 percent degradation of efficiency due to the cooling.

The first of the three mechanisms is hardest to quantify, and we must rely mainly on experimental data. In an experiment reported in reference 6.8 the kinetic energy efficiency was measured for a nozzle cascade with injection through each of 12 rows of holes, as shown in figure 6.14. The kinetic energy efficiency is here defined as

$$\eta_{KE} = \frac{\text{actual exit kinetic energy}}{\text{ideal exit kinetic energy for main flow plus coolant flow}};$$

the data plotted were selected from reference 6.8 so that the stagnation pressure of the cooling airflow in the blade core equalled the stagnation pressure of the main flow. The change in η_{KE} divided by the fractional cooling flow ε is plotted for the points on both the suction and pressure surfaces. The pressure distributions on these surfaces are also shown. There was about 0.2 percent reduction in η_{KE} for each percent cooling flow on the pressure side, and about 0.5 percent on the suction side.

Thus to the 2.7 percent loss per percent of cooling flow estimated from (6.19), we should add about another 0.5 percent, giving a total of some 3.2 percent loss in turbine efficiency per percent of cooling flow.

There are few systematic data in the literature against which to test an estimate such as this. But recently one comparison has been published of the efficiency of a cooled turbine to an uncooled turbine of the same design

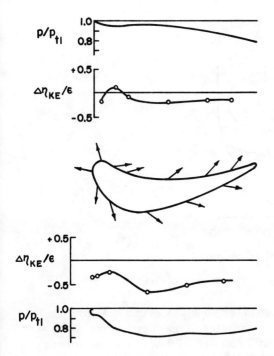

6.14   Effect of film-cooling airflow on kinetic energy efficiency of turbine cascade. The change in η_{KE} is for each row of cooling holes flowing alone (adapted from reference 6.8).

(reference 6.9). The conclusion can be drawn from the data that a 2.5 percent rotor cooling flow reduced the efficiency about 6.5 percent.

6.4   Turbine Similarity

The representation of empirical turbine performance in terms of corrected parameters is entirely analogous to that of compressors discussed in section 5.4. Tangential Mach number is represented by the corrected speed, $N/\sqrt{\theta}$, where θ is the inlet stagnation temperature divided by the reference standard and N is rpm for a given turbine. Axial Mach number is represented by the corrected weight flow $W\sqrt{\theta}/\delta$.

A map for a typical 50 percent reaction (at mid radius) single-stage turbine is shown in this format in figure 6.15. The abscissa has been taken as $(W\sqrt{\theta}/\delta)(N/\sqrt{\theta})$ instead of as $W\sqrt{\theta}/\delta$ because the mass flow is nearly independent of speed for $1/\pi_t > 2.5$; thus all speed characteristics collapse onto a single line, and the turbine has the mass flow characteristic of a choked nozzle. A separate plot would then be needed to show η_t as a function of

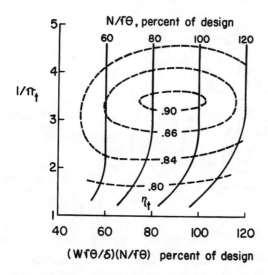

6.15   Typical turbine performance map.

$N/\sqrt{\theta}$. The choking does not occur in the same way as in a simple nozzle because of the energy extraction by the rotor. It may occur at the nozzle exit or at the rotor blade exit, or it may be a result of flow limitations at both positions. If the nozzles choke with increasing $1/\pi_t$ without choking in the blades, then we find the characteristic shown in figure 6.15, that $W\sqrt{\theta}/\delta$ is independent of $N/\sqrt{\theta}$ for large $1/\pi_t$. If, on the other hand, the rotor outlet or exit annulus were choked, $W\sqrt{\theta}/\delta$ would depend on $N/\sqrt{\theta}$ for a given $1/\pi_t$ because changes in rotor speed would affect the stagnation pressure and temperature, hence $(\rho u)^*$, in the downstream portions of the turbine. We see from figure 6.15 that for low $1/\pi_t$ the mass flow decreases with increasing $N/\sqrt{\theta}$ for fixed $1/\pi_t$. This is because the stagnation pressure drop across the rotor is larger at larger $N/\sqrt{\theta}$ ($1 - \tau_t \propto M_T^2$), so for given upstream and downstream pressures, there is less pressure drop across the nozzles at large $N/\sqrt{\theta}$, hence less mass flow.

Because η_t does not vary as rapidly with off-design variations as in a compressor, the turbine characteristic can be approximated for preliminary design calculations by a constant η_t and a choked mass flow characteristic, so that $W\sqrt{\theta}/\delta = $ const, $\eta_t = $ const, is an approximate description of the turbine characteristics for purposes of engine performance calculation.

References

6.1 J. H. Horlock, *Axial Flow Turbines*, Kruger Publishing Co., 1973.

6.2 W. L. Stewart, W. J. Whitney, and R. Y. Wong, "A Study of Boundary Layer Characteristics of Turbomachine Blade Rows and Their Relation to Overall Blade Loss," *Journal of Basic Engineering*, Vol. 82-D (1960), p. 588.

6.3 D. G. Ainley and G. C. R. Mathieson, "A Method of Performance Estimation for Axial Flow Turbines," ARC R & M No. 2974, Her Majesty's Stationery Office, 1957.

6.4 R. D. Lander, R. W. Fish, M. Suo, L. L. Grimme, and J. E. Muska, AIAA Paper No. 72-9, 1972.

6.5 M. E. Crawford, H. Choe, W. M. Kays, and R. J. Moffatt, "Full Coverage Film Cooling Heat Transfer Studies—A Summary of the Data for Normal-Hole Injection and 30° Slant-Hole Injection," Rept. HMT-19, Stanford University, 1975.

6.6 J. F. Louis, A. M. Demerjian, G. N. Goulios, R. F. Topping and J. M. Wiedhopf, "Short Duration Studies of Turbine Heat Transfer and Film Cooling Effectiveness," ASME Paper 74-GT-113, June 1974. Also: A. M. Demerjian, "An Analytical and Experimental Investigation of Film Cooling Effectiveness Over a Flat Plate," Ph. D. thesis, MIT, 1975.

6.7 S. N. Suciu, "High Temperature Turbine Design Considerations," SAE Paper No. 710462, 1971.

6.8 Herman W. Prust, Jr., "Two-Dimensional Cold-Air Cascade Study of a Film-Cooled Turbine Stator Blade. II, Experimental Results of Full Film Cooling Tests," NASA TM x–3153, 1975.

6.9 H. Nouse et al., "Experimental Results of Full Scale Air-Cooled Turbine Tests," ASME Paper No. 75–GT–116, April 1975.

Problems

6.1 The analogue of the centrifugal compressor is the radial inflow turbine. Supposing that such a device has radial vanes and zero exit swirl, derive an expression for its stagnation temperature ratio as a function of tip tangential Mach number. What is its effective degree of reaction?

6.2 A turbine is to be "designed" to drive a compressor that has the following characteristics at its design point:

Weight flow	100 kg s^{-1}
Pressure ratio	8
Efficiency	0.85
Tip speed	400 m s^{-1}
Tip diameter	0.7 m
Inlet conditions	1 atm, 300°K

The turbine inlet temperature is to be 1400°K.

There are a number of design options. You might consider

a. zero exit swirl from rotor,

b. 50 percent reaction ($R = 0.5$) at midspan (gives M_T at midspan),

c. $M_b = 1$ at midspan (gives M_a).

Sketch the velocity triangles at hub, midspan, and tip, and the shape of the annulus. Also, estimate the blade root stress assuming the blades are of constant cross-section, and calculate the *horsepower* output of this turbine.

6.3 Using flat plate boundary layer relations for heat transfer and a free-stream Mach number of 1, estimate the heat transfer to the surface of a turbine nozzle vane kept at a uniform surface temperature of 1200°K by internal cooling when the turbine inlet pressure and temperature are 25 atm and 1600°K. Plot q_w as a function of distance from the leading edge for a blade with 4 cm chord. Take $\mu = 0.6 \times 10^{-4}$ kg/m s. If the thermal conductivity $k = 0.4$ watt/cm °K, how thick can the blade's skin be if the temperature drop in the skin is not to exceed 100°K?

6.4 Using the data of figure 6.11, find the factor by which the heat flux to a wall is reduced by film cooling from a set of two rows of holes, at a distance of 10 equivalent slot widths from the holes, for a mass blowing parameter $m = 0.43$, when the adiabatic wall temperature $T_r = 1600$°K, the wall temperature $T_w = 1200$°K, and the coolant temperature $T_c = 1000$°K.

6.5 To design (approximately) the nozzle-vane cooling for a turbine proceed as follows:

1. Estimate, using flat plate relations for boundary layers, the heat transfer rate for an internally cooled blade, taking the metal temperature to be uniform. Assume laminar to turbulent transition occurs at some Re_x on the order of 2–5 $\times$ 10^5.

2. Calculate the internal convection cooling airflow required to cool the trailing edge, taking reasonable thicknesses for the blade walls and cooling passage. Model the blade by a two-dimensional structure for this purpose, and use (6.10) for the internal flow.

3. Now calculate the amount of cooling this internal airflow will give the rest of the blade surface, that is, find q for a reasonable thickness.

4. Using the film cooling correlation of figure 6.11 or 6.13, find the distribution of film cooling holes and air needed to cool the "hotspots" on the blade, for example, at the leading edge and at transition.

5. Make a sketch of the vane cross-section.

Carry out this procedure for a vane with a chord of 4 cm, with $T_r = 1600$°K, $T_w = 1200$°K, $T_c = 1000$°K, $p_{t4} = 25$ atm. Take $k = 40$ watts/m °K and $\mu \approx 0.6 \times 10^{-4}$ kg/m s.

6.6 Equation (6.6) shows that the rotor-relative temperature increases with the degree of reaction R, for fixed M_T and M_b. Suppose, however, that the temperature ratio τ_t is held constant as R is varied. Then how does T_{tr}/T_{ta} vary with R? Does your result imply that increasing R relieves the high-temperature, high-stress problem of the turbine rotor?

7
Structure of Engine Turbomachinery

The requirements for operation at high tangential velocities and in the case of the turbine at high temperatures impose severe constraints on the design of turbomachinery. These constraints must be understood at least qualitatively to appreciate the compromises required in engine design. In the following sections, some of the more important considerations will be discussed in a qualitative way. Techniques are available for much more precise treatment of all these matters, some in the literature and some as part of the fund of proprietary information held by each engine manufacturer.

7.1 Centrifugal Stresses

A brief introduction to the centrifugal stress problem was given in section 1.10 where the stress in a rotating bar of constant cross section was computed. Now consider a rotor, such as sketched in figure 7.1. It consists of a disc of variable thickness, with inner and outer rims and blades attached to the outer rim. The reason for variation of the disc's thickness will become clear as we proceed. An analysis of this structure for arbitrary shape can be done within the framework of the theory of elasticity, but to simplify the argument we will make some assumptions.

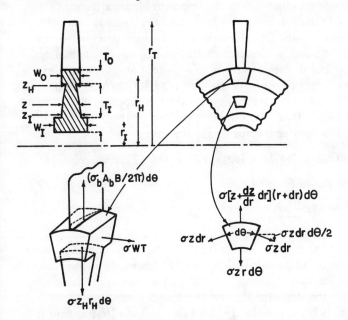

7.1   Schematic of turbomachine rotor, consisting of disc and blades, with free-body diagram showing stresses.

7.1.1 Discs

Consider first the disc, assuming that its (axial) thickness can be so adjusted that the tensile stress is uniform in the disc and the same in all directions in planes perpendicular to the axis of rotation. Let this stress be σ and consider the force balance on a small element of the disc as shown in figure 7.1. The centrifugal force on the element is $\rho\omega^2 r \, d\theta \, dr \, z$, and this is opposed by the difference between the radial forces acting on the inner and outer surfaces of the element, plus the radial component of the tangential forces acting on the two sides. The latter force is $\sigma z \, dr \sin(d\theta) \approx \sigma z \, dr \, d\theta$, so that

$$\rho z \omega^2 r^2 \, d\theta \, dr = \sigma z r \, d\theta - \sigma(r + dr)[z + (dz/dr) \, dr] \, d\theta + \sigma z \, dr \, d\theta$$

$$= -\sigma r (dz/dr) \, dr \, d\theta$$

and the thickness distribution in r which gives constant σ is

$$(1/z)(dz/dr) = -(\rho\omega^2 r)/\sigma \quad \text{or} \quad z = \text{const } e^{-(\rho\omega^2 r^2)/2\sigma} \tag{7.1}$$

The thickness decreases toward the edge of the disc, and we see that σ can be made as small as we like; but the smaller σ, the larger the variation of z with r.

Next suppose we add an outer rim to the disc with blades attached to it as shown in figure 7.1; the rim has a width W_0 and thickness T_0, and a tangential stress σ equal to the disc stress, so that its circumferential strain will be the same. If we imagine the force exerted on the rim by the blades to be uniformly distributed, then for B blades with root area A_b and stress σ_b, the force on an element of rim with angle $d\theta$ is $(\sigma_b B A_b / 2\pi) \, d\theta$. The force exerted on the rim by the disc is $\sigma z_H r_H \, d\theta$, and the rim supports itself through the circumferential force $\sigma W_0 T_0 \, d\theta$. Equating the sum of these forces in the radial direction to the centrifugal force on the rim gives

$$(\rho W_0 \omega^2 r_H^2 T_0) \, d\theta + (\sigma_b A_b B/2\pi) \, d\theta = (\sigma z_H r_H) \, d\theta + (\sigma W_0 T_0) \, d\theta;$$

this equation can be solved for σ to give

$$\sigma = \frac{\rho\omega^2 r_H^2 + (B/2\pi)\sigma_b(A_b/W_0 T_0)}{1 + (z_H r_H/W_0 T_0)}.$$

This result shows that the stress has a contribution from the centrifugal force in the rim itself, $\rho\omega^2 r_H^2$, and a contribution from the blades. It further shows that the disc *reduces* the stress level, as $z_H r_H/W_0 T_0$ becomes appreciable compared to 1, below the stress for the rim and blades alone.

The blade stress σ_b is related to the tip speed ωr_T and to r_H/r_T, as shown in section 1.10. If the blade has constant cross section with radius, then $\sigma_b = (\rho\omega^2 r_T^2/2)[1 - (r_H/r_T)^2]$, so we can write the stress as

$$(\sigma/\rho\omega^2 r_T^2) = \frac{(r_H/r_T)^2 + (B/4\pi)(A_b/W_0T_0)[1 - (r_H/r_T)^2]}{1 + (z_H r_H/W_0 T_0)}. \tag{7.2}$$

A similar argument applies to the inner rim, except that in this case the rim supports the disc, so that

$$(\sigma/\rho\omega^2 r_T^2) = (r_I/r_T)^2/\{1 - [(z_I r_I)/(W_I T_I)]\}. \tag{7.3}$$

The shape of the disc between the two rims is found by evaluating the constant in (7.1), say, at the edge near the outer rim. Putting z there equal to z_H gives

$$z/z_H = \exp\{(\rho\omega^2 r_T^2/2\sigma)[(r_H/r_T)^2 - (r/r_T)^2]\}. \tag{7.4}$$

A schematic procedure for choosing the disc shape can now be seen as follows. Suppose first that the following are known in (7.2):

the permissible stress σ;

the desired tip speed ωr_T;

the hub and tip radii r_H and r_T; and

the number B and the root area A_b of the blades and their chord.
The quantities in (7.2) still to be determined are $W_0 T_0$ and z_H. If we choose a rim width and thickness, then the disc thickness is determined by this equation. Going then to (7.4), if we choose r_I/r_T, z_I/z_H is determined; finally from (7.3), $W_I T_I$ is found.

To illustrate this procedure, consider the following example: $\sigma/\rho\omega^2 r_T^2 = 0.5$, $r_H/r_T = 0.5$, $B = 40$, $A_b/W_0 T_0 = 0.2$, $r_I/r_T = 0.1$. (7.2) requires $z_H r_H/W_0 T_0 = 1.91$. For the sake of simplicity suppose $W_0 = T_0$, and assume $W_0 = 0.1 r_T$. Then $z_H = 0.038 r_T$. From (7.4), $z_I = 1.27 z_H$, and from (7.3), $W_I/r_T = 0.12$. This disc is shown in cross-section in figure 7.2 by the full lines. Serious stress concentrations would result from the sudden area changes at the junctures of disc and rims, so they would in practice be faired as shown by the dashed lines.

Some general observations can be derived from this simple analysis. First, geometrically similar rotors of varying size will have similar stress distributions if $\sigma/\rho\omega^2 r_T^2$ is constant with changing size. Other things being equal, the disc stress level increases as the hub/tip radius ratio decreases, because the second term in the numerator of (7.2), representing the blade forces, increases. This term contributed about 0.65 of the total stress for the above example.

An important criterion for design of gas turbine discs is the "burst speed," the speed at which the disc will separate into pieces and fly apart. The fragments have so much energy that in such an event total wreckage of a gas turbine and loss of an aircraft is likely to result. A fairly reliable estimate of the burst speed is obtained by regarding each of two halves of the

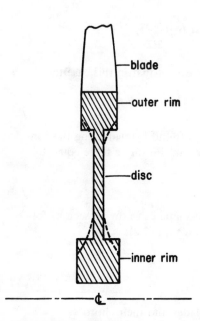

7.2   Sketch of the disc design described in section 7.1.1, with inner and outer rims and tapered disc.

disc (separated by a diameter) as free bodies held together by a uniform stress. The burst speed is that for which this uniform stress equals the material's ultimate strength. This relatively simple model applies because the disc material is ductile enough to yield before fracturing, thus distributing the stress uniformly. Typically, the burst speed should be 1.3 to 1.4 times design speed.

7.1.2   Centrifugal Stresses in Blades

The simple argument given in section 1.10 is readily generalized to account for variations in blade cross-sectional area with radius. Letting $A(r)$ be the area, the mean stress in the cross-section at any radius r will be

$$\sigma A(r) = \int_{r}^{r_T} \rho \omega^2 A(r) r \, dr.$$

For a linear taper from hub to tip, that is, $A(r) = A_H(1 - \alpha r)$, we find

$$\sigma(r_H)/\rho\omega^2 r_T^2 = \tfrac{1}{2}\{1 - (r_H/r_T)^2 - (2\alpha r_T/3)[1 - (r_H/r_T)^3]\}. \tag{7.5}$$

If, for example, $r_H/r_T = 0.5$ and $\alpha r_T = 0.5$, the taper reduces the right-hand side from 0.75 to 0.46, a 40 percent decrease in σ, for given ωr_T.

For untapered blades ($\alpha = 0$), (7.5) shows that the stress level is proportional to the ratio of annulus flow area to total area, so it is clear that the highest blade stress levels will occur in the first stage of a compressor and in the last stage of a turbine.

7.2 Gas Bending Loads on Blades

The gas dynamic forces acting on a small radial element of a rotating blade are indicated in figure 7.3 to be principally axial and tangential. There is also a radial component, but it is always negligible compared to the centrifugal forces. The axial and tangential forces generate a bending moment about the root of the blade, which may be thought of as a cantilevered beam. The maximum stress due to the gas forces will be at the root; since the centrifugal stresses are largest there also, it is the critical location.

To estimate the stresses, we can calculate the moment M_z about an axis through the blade root and parallel to the rotational axis and the moment M_θ about a tangential axis, then combine them to find the moments about the principal axes of the blade cross-section and hence the bending stress.

For the sake of simplicity let us assume that the fluid streamlines do not change radius as they pass through the rotor, that the fluid is incompressible, and that the axial velocity is constant. The tangential force acting on the element dr of the blade is then simply $dF_\theta = \dot{m}(v_c - v_b)$ where $\dot{m}$ is the mass flow associated with the element dr of the blade, and $v_c - v_b$ is the tangential

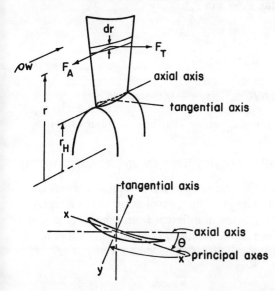

7.3   Diagram of tangential and axial forces acting on a blade to produce bending stresses at the root about the axial and tangential directions, or about the principal axes x, y.

velocity change across the rotor. If B is the number of blades, $B\, dF_\theta = \rho w 2\pi r(v_c - v_b)\, dr$. The axial force is due to the static pressure rise across the blade row, since we are assuming that the axial velocity is constant. It is therefore $B\, dF_z = (p_c - p_b)(2\pi r\, dr)$. The two moments are then

$$BM_z = \int_{r_H}^{r_T} B\, dF_\theta(r - r_H) = \int_{r_H}^{r_T} \rho w 2\pi r(v_c - v_b)(r - r_H)\, dr,$$

$$BM_\theta = \int_{r_H}^{r_T} B\, dF_z(r - r_H) = \int_{r_H}^{r_T} (p_c - p_b)2\pi r(r - r_H)\, dr.$$

To estimate the first integrand in terms of easily interpreted quantities, we may use the Euler equation (5.7), which states that $v_c - v_b = c_p(T_{tc} - T_{tb})/\omega r = c_p T_{tb}(\tau_s - 1)/\omega r$. The pressure difference can be found by noting that the stagnation pressure of the fluid *relative to the rotor* is constant across the rotor, so $p_c - p_b = \rho[(V_b')^2 - (V_c')^2]/2$ in the incompressible limit, and this can be written

$$p_c - p_b = (\rho w^2/2)(\tan^2 \beta_b' - \tan^2 \beta_c')$$
$$= (\rho w^2/2)(\tan \beta_b' - \tan \beta_c')(\tan \beta_b' + \tan \beta_c').$$

But for constant axial velocity, $\tan \beta_b' - \tan \beta_c' = (v_c - v_b)/w$, so that

$$p_c - p_b = \rho w(v_c - v_b)(\tan \beta_b' + \tan \beta_c')/2$$
$$= [c_p T_{tb}(\tau_s - 1)/\omega r][\rho w(\tan \beta_b' + \tan \beta_c')/2].$$

Finally, the moments are

$$BM_z = \int_{r_H}^{r_T} (1/\omega)[2\pi \rho w c_p T_{tb}(\tau_s - 1)](r - r_H)\, dr,$$

$$BM_\theta = \int_{r_H}^{r_T} (1/\omega)[\pi \rho w(\tan \beta_b' + \tan \beta_c')c_p T_{tb}(\tau_s - 1)](r - r_H)\, dr.$$

If τ_s is independent of r, the first integral can be evaluated without further assumptions; it becomes

$$BM_z = (1/\omega)\rho w c_p T_{tb}(\tau_s - 1)(\pi r_T^2)\{[1 - (r_H/r_T)^2]\{1 - [2r_H/(r_T + r_H)]\}\}. \qquad (7.6)$$

The second integrand is modified by the factor $(\tan \beta_b' + \tan \beta_c')/2$, which is just the mean stagger angle. Because most of the contribution to the integral comes from the tip, where $\tan \beta_c'$ is not much different from $\tan \beta_b'$, we approximate the factor by $\tan \beta_b'$. Then in the notation of section 5.1.2, $\tan \beta_b' = [(\omega r_T/w) - A](r/r_T)$, where A expresses the inlet guide vane turning. The tangential moment is then

$$BM_\theta = (1/\omega)\rho w c_p T_{tb}(\tau_s - 1)(\pi r_T^2)$$
$$\cdot \{[(\omega r_T/w) - A][(2/3) - (r_H/r_T) + (1/3)(r_H/r_T)^3]\}. \qquad (7.7)$$

The moments about the principal axes of the blade cross-section are related to M_z and M_θ by

$$M_{xx} = M_z \cos\theta + M_\theta \sin\theta, \qquad M_{yy} = M_\theta \cos\theta - M_z \sin\theta, \qquad (7.8)$$

and the maximum bending stress is then

$$\sigma_{\text{bend}} = [(M_{xx} y_{\text{max}})/I_{xx}] + [(M_{yy} x_{\text{max}})/I_{yy}] \qquad (7.9)$$

where y_{max} and x_{max} are the maximum distances from the axes and I_{xx} and I_{yy} are the principal moments of inertia.

To estimate the magnitude of the bending stresses, consider just the factor in front of the large braces in (7.6) and (7.7); the factors involving r_H/r_T are of order unity. Take $\sigma_{\text{bend}} \approx My/I$; $y \approx t$, the blade thickness, and $I \approx ct^3$, where c is the blade chord; then we find that

$$\sigma/p \approx (w/\omega r_T)(\tau_s - 1)(s/2c)(r_T/t)^2 \qquad (7.10)$$

apart from a numerical factor of order unity. This result shows a number of important scaling rules. The stress is proportional to the air pressure and to the temperature rise of the blade row. It is inversely proportional to the solidity and inversely proportional to the *square* of the ratio of blade thickness to tip radius. Taking for example $w/\omega r_T = 1/2$, $\tau_s - 1 = 1/4$, $c/s = 2$, $r_T/t = (r_T/c)(c/t) = (6)(10) = 60$ we find $\sigma/p = 110$, and for $p = 1$ atm, $\sigma \approx 116$ atm. This is a modest stress level, but for thinner blades and higher aspect ratios $[(r_T - r_H)/c]$, the bending stress can become an appreciable part of the total stress.

7.3 Thermal Stresses

Thermal stresses result from the tendency of materials to expand with increasing temperatures. In a turbine blade, for example, if the leading and trailing edges are hotter than the middle part of the airfoil, they tend to grow radially. But to grow they must stretch the center of the blade. The result is a compressive stress in the leading and trailing edges and tension in the center. This tensile stress would add to the centrifugal stress already present.

The thermal stress distribution can be analyzed as follows. Suppose first that the blade cross-section were divided into a large number of small elements, as suggested in figure 7.4, all independent of each other mechanically, so each could expand radially in accordance with its temperature and centrifugal stress. Suppose also that the temperature distribution in the blade were known. The total radial strain (fractional elongation) would then be

$$\varepsilon = (\sigma_c/E) + \alpha(T - \overline{T}) \qquad (7.11)$$

elements hotter, cooler than average

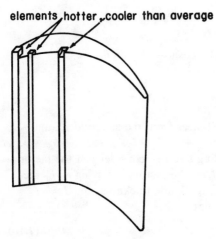

7.4 Schematic representation of the mechanism that generates thermal stresses in blades due to chordwise temperature gradient.

where α is the thermal coefficient of expansion, σ_c is the centrifugal stress, E is the elastic modulus of the material, and $\overline{T}$ is the area-averaged temperature across the blade cross-section. If the blade is long and has the assumed temperature distribution all along its length, then the total strain ε must be the same in each of the elements. The thermal stress is the stress that must be added to each individual fiber so that its strain equals the average strain. That is, if $\bar{\varepsilon}$ is the average strain,

$$\sigma_T/E = \bar{\varepsilon} - (\sigma_c/E) - \alpha(T - \overline{T}).$$

Since the thermal stresses are all within the blade, they must sum to zero over the cross-section, so that $\bar{\varepsilon} = \sigma_c/E$, and the thermal stress is just $\sigma_T/E = -\alpha(T - \overline{T})$. The *total* stress is

$$\sigma/E = (\sigma_c/E) + \alpha(T - \overline{T}). \tag{7.12}$$

The centrifugal stress, being tensile, is negative, so (7.12) indicates that the tensile stress is highest in the lower temperature elements. A typical value for α is about $0.9 \times 10^{-5} \,^\circ\mathrm{K}^{-1}$ while E is about 2×10^6 atm for steels, so αE is 18 atm $^\circ\mathrm{K}^{-1}$, and we see that a temperature difference of $100\,^\circ\mathrm{K}$ will produce a thermal stress of 1800 atm!

A better appreciation for the importance of thermal stresses can be had by relating the temperature difference to heat fluxes required in internally cooled blades. Consider the skin of thickness t of the blade in figure 6.9. For this case, $q_w = k(\Delta T/t)$ where ΔT is the temperature drop across the skin. If we put $T - \overline{T}$ in (7.12) equal to $\Delta T/2$, then $\sigma - \sigma_c = \alpha E(\Delta T/2) = \alpha E q_w t/2k$.

Recall from (6.8) that $q_w \approx \rho u c_p (T_r - T_w) St$. Combining these facts leads to

$$(\sigma - \sigma_c)/p = (\alpha E/k)(ut)[\gamma/2(\gamma - 1)][1 + \tfrac{1}{2}(\gamma - 1)M_{tr}^2][1 - (T_w/T_r)]St. \qquad (7.13)$$

The first factor, $\alpha E/k$, expresses the material's susceptibility to thermal stresses. For high alloy steels it is of order $54,000$ s m^{-2}. The second factor is the flow velocity over the blade times the thickness of the blade skin; for consistency it would have dimensions of m^2 s^{-1}. As an example, suppose $u = 600$ m s^{-1}, $t = 1.5$ mm, $St = 0.003$; then $(\sigma - \sigma_c)/p \approx 300[1 - (T_w/T_r)]$. In a high pressure ratio engine, $p \approx 20$ atm, so $\sigma - \sigma_c \approx 6000[1 - (T_w/T_r)]$ atm. Thus, cooling just 10 percent below the gas temperature gives a thermal stress of 600 atm under such circumstances in a 1.5 mm thick skin. In other words, increasing T_r increases the blade thermal stress, other things being equal. The thermal stress for the above example is 3000 atm where T_r is twice T_w. This is one reason that film or aspiration cooling must be used for turbines with large ratios of gas to metal temperature. From the thermal stress standpoint, the blade thickness must be maintained small as engine size is increased; this is another example of the inapplicability of geometric scaling in going from small to large engines.

The importance of precision in cooling designs can readily be appreciated from these estimates. With a heat transfer distribution such as in figure 6.8, the internal cooling and film cooling must be arranged to make the metal temperature T_w as nearly uniform as possible, at a level acceptable to the material (currently about 1250°K maximum for aircraft applications and nearer 1150°K for ground applications). A difference of perhaps 15°K in local temperature can result in a factor of two change in turbine life at the 1250°K level. Now if the blade is held at 1250°K in a 1600°K stream, with cooling air at 850°K, the adiabatic effectiveness must be about 0.5, and it must be known within about 5 percent to hold the temperature variations to 15°K. Currently, the limit on turbine inlet temperature is set by the precision of blade-cooling design techniques.

There can be serious thermal stresses in turbine discs also. When hot gases impinge on the outer rim of the disc or when heat is conducted into it from the blades, it tends to expand relative to the cooler inner portions of the disc. This reduces the tensile stress in the rim and increases that in the disc, so that care must be exercised in cooling the disc and blade mounts.

7.4 Critical Speeds and Vibration

In figure 7.5, a gas turbine engine is idealized for the purpose of vibration analysis as a flexible rotor consisting of a lumped mass mounted on a flexible shaft mounted in flexible bearings. This model illustrates some of the more important vibrational characteristics of engines. In addition, the individual

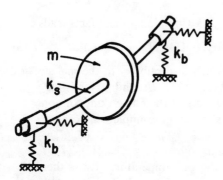

7.5   Model of a turbomachine rotor of mass m, with shaft spring constant k_s and bearing spring constant k_b.

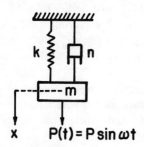

7.6   Single degree of freedom mass-spring system with damping.

blades and discs can vibrate both alone and cooperatively, but different models will be required to represent these modes.

Frequent reference will be made to the forced oscillation of a single degree of freedom mass-spring system such as that sketched in figure 7.6. The differential equation describing its motion for a harmonic driving force is (reference 7.1)

$$\ddot{x} + 2n\dot{x} + \omega_n^2 x = (P/m) \sin \omega t \tag{7.14}$$

where $\omega_n^2 = k/m$ is the natural frequency, n is a damping coefficient, and P is the amplitude of the exciting force. The solution for the forced steady-state vibration is $x = A \sin (\omega t - \phi)$ where the amplitude A is given by

$$A = (P/k)\{[1 - (\omega^2/\omega_n^2)]^2 + [(4n^2\omega^2)/\omega_n^4]\}^{-1/2} \tag{7.15}$$

and the phase angle ϕ is given by

$$\tan \phi = 2n\omega/(\omega_n^2 - \omega^2). \tag{7.16}$$

Since P/k is the static deflection under the load P, the square root factor is

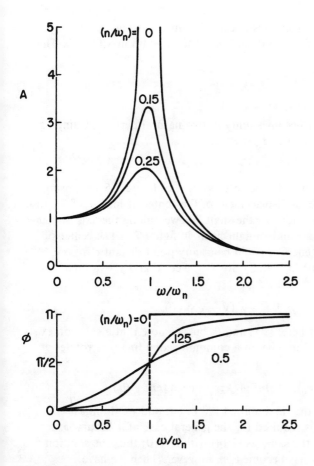

7.7 Amplification factor A and phase lag ϕ for single degree of freedom vibrating system of figure 7.6.

the ratio by which the static deflection is amplified dynamically. For a system with no damping it would become infinite at resonance, that is, for $\omega = \omega_n$. The amplification is shown as a function of ω/ω_n and n/ω_n in figure 7.7 as is the phase angle ϕ. We see that as ω/ω_n increases from zero, ϕ increases; the motion lags the forcing, reaching a lag of $\pi/2$ at resonance and approaching π as $\omega/\omega_n \to \infty$.

The behavior near resonance is critical for gas turbines since they must at times pass through resonant speeds in reaching their normal operating speed. This steady-state solution indicates a very large amplification will occur at resonance for low damping, but this large amplitude does not occur instantly when the system is driven at resonant frequency, because it is initially at rest. The amplitude in fact increases linearly in time; this may be seen by constructing the solution for $n = 0$ with the initial condition $x(0) = 0$. The result is

$$x = -\frac{P}{k}\left(\frac{\omega_n t}{2}\right)\cos \omega_n t, \tag{7.17}$$

so the amplification factor becomes unity in two natural periods. Damping would of course reduce the rate of growth.

7.4.1 Shaft Critical Speed

Let us now apply these results to the rotor system of figure 7.5, supposing that excitation is caused by a displacement of the center of mass c of the disc by a distance e from the center s of the shaft. Viewed along the axis of rotation, the relation of points c and s might be as in figure 7.8. The center of mass c is at x, y, and in general s could be at any point distant e from c. The deflections of the shaft in x and y are $x - e \cos \theta$ and $y - e \sin \theta$; thus the two equations of motion are

$$m\ddot{x} = -k_s(x - e \cos \theta), \qquad m\ddot{y} = -k_s(y - e \sin \theta). \tag{7.18}$$

There are three unknowns here, x, y, and θ, but we note that a solution exists for $\theta = \theta_0 + \omega t$, where s and c lie on a common radial line that rotates at angular velocity ω. Then

$$m\ddot{x} + k_s x = e \cos (\omega t + \theta_0), \qquad m\ddot{y} + k_s y = e \sin (\omega t + \theta_0), \tag{7.19}$$

so the motion in y is just that in x shifted $\pi/2$ in ωt, and we do not need the third equation which is required for the general case of arbitrary ϕ. Either of the equations is the same as (7.14) with $n = 0$; thus the solution has the same form where P/m has become e/m, however. Then we have

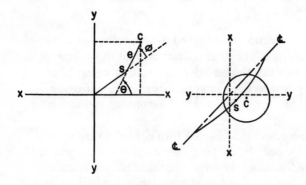

7.8   Relation of center of mass c to axis of rotation $x = 0$, $y = 0$ and centerline deflected shaft s for disc-shaft assembly.

$$x = \frac{e}{1 - (\omega/\omega_{ns})^2} \cos \omega t$$

where $w_{ns}^2 = k_s/m$ and the phase angle ϕ is zero for $\omega < \omega_{ns}$, but π for $\omega > \omega_{ns}$. That is, for $\omega < \omega_{ns}$ the center of mass lies outside the shaft center, and as $\omega \to \omega_n$ from below, x/e tends toward ∞. But for $\omega > \omega_{ns}$ the center of mass lies inside the shaft center and $x \to -0$ as $\omega/\omega_{ns} \to \infty$, so the disc tends to rotate about its center of mass. The first condition is termed *below critical speed*, the second *above critical speed*. Note that the critical *angular speed* of the rotor is just equal to the vibrational angular frequency in transverse vibrations it would exhibit if not rotating.

There are three points of practical importance here. First, the amplitude of vibration becomes large near critical speed. Second, there is little damping because the geometry of the rotor is fixed as it spins in this mode. Thus if a rotor is to be run above critical speed, the passage through critical speed must be rapid. Finally, the rotor can be run above critical speed with a small eccentricity; the deflection of the shaft approaches a limiting value e as the speed becomes large.

Many gas turbine rotors do in fact run above shaft critical. But the modern tendency is to design rotors to be very stiff in bending by using large diameter conical shafts such as that sketched in figure 6.7. A number of examples can be seen in figures 1.15 to 1.19 as well.

7.4.2 Structural Loads and Critical Speeds

No rotor is perfectly balanced, so the rotation will always lead to some shaft deflection and to a fluctuating load on the bearings. For the system of figure 7.5, with $k_b \gg k_s$, the load would be half the spring force of the shaft per bearing, or $-k_s(x - e \cos \omega t)$ for the two bearings. With the expression for x, then,

$$P(t) = -k_s e \left[\frac{(\omega/\omega_{ns})^2}{1 - (\omega/\omega_{ns})^2} \right] \cos \omega t.$$

The oscillating load on the bearings first increases as ω^2, tending to infinity as $\omega \to \omega_{ns}$, then switches sign and decreases from infinity, approaching $k_s e \cos \omega t$ as $\omega/\omega_{ns} \to \infty$. Thus the "flexible shaft" rotor will have moderate bearing loads at high rotor speeds, but a stiff rotor would produce very large loads in rigid bearings. This problem is overcome by making the bearing mounts flexible. If we think of the whole rotor system for $\omega_{ns} \gg \omega$ as a mass mounted on a spring k_b representing the bearing stiffness, the assembly is directly analogous to the mass-spring system of figure 7.6 and will have the same response to the excitation which is now $P(t) = -me\omega^2 \cos \omega t$. The motion of the shaft at the bearing will be $x_b = A_b \cos \omega t$ where

$$A_b = -e/\{[(\omega_{nb}^2/\omega^2) - 1]^2 + [4n_b^2/\omega^2]\}^{1/2};$$ (7.20)

this shows that $A_b/e \to 0$ for $(\omega_{nb}/\omega)^2 \to \infty$, as expected; but further for $(\omega_{nb}/\omega)^2 \to 0$ (for very soft bearings), $A_b/e \to 1/[1 + (4n_b^2/\omega^2)]^{1/2}$, which is always less than 1. The force transmitted to the engine structure through the bearing will be

$$P_b(t) = x_b k_b = x_b m \omega_{nb}^2 = em\omega_{nb}^2/[1 + (4n_b^2/\omega^2)]^{1/2}.$$ (7.21)

This result shows that reducing the stiffness and increasing the damping of the bearing mounts will greatly reduce the vibrational loads on the engine structure without resulting in shaft displacements in excess of e provided $\omega^2 \gg \omega_{nb}^2$.

Most modern aircraft gas turbines use very soft oil damped bearings in which the outer race is actually supported in its housing by pressurized oil.

7.4.3 Blade and Disc Vibration
Each of the turbomachine blades individually and the assembly of blades and their supporting disc collectively are capable of vibrating in a number of modes. If the blades are attached only at their roots, then in first approximation they may be thought of as cantilevered beams, as sketched in figure 7.9. Such a blade can vibrate in bending and in torsion. The density of most turbomachine blades is so large relative to that of the fluid that the modes of vibration are very nearly those for the blade in vacuum. That is, the effect of the fluid is only to provide excitation or damping; it does not change the mode shape or frequency. The frequencies are influenced by temperature through its effect on the elastic modulus and by the speed of rotation N, the centrifugal force raising the frequency. The possibilities for excitation of a given blade can be represented by a "Campbell diagram" in which the frequencies of the various modes are plotted versus N, as in figure 7.10, and lines representing multiples of N are superimposed. Any intersection of the latter with a blade frequency curve within the normal speed range of the machine is a potential source of destructive vibration. Not all can be avoided of course. The strongest excitation usually results from resonances with N, $2N$, $3N$, and perhaps $4N$ excitations, so it is usual to avoid these.

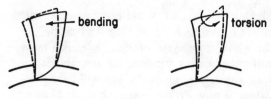

7.9 Illustrating first bending and torsional modes of vibration of a cantilevered compressor blade.

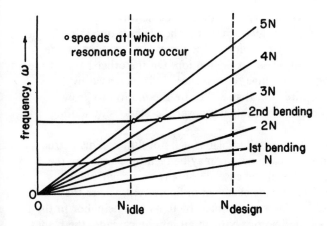

7.10 Frequency diagram or "Campbell diagram" representing possible resonant excitation points for a blade in bending vibration.

Often blades are coupled at their tips or at an intermediate radius by "shrouds" which at operating speed effectively join into a solid ring. Then more complex blade-shroud modes are possible (reference 7.2).

The discs by themselves are capable of vibration in various plate-type modes. In the "umbrella" mode, the disc center oscillates axially against the rim. A series of modes occur in which the nodes are diameters, and the disc rim assumes a wavy axial displacement between nodes. These are particularly likely to be excited by a cooperative bending vibration of the blades.

7.5 Blade Flutter

The vibrations discussed in section 7.4 all result from some *forced excitation* resulting from the rotation of the machine, either due to structural inertia forces or to fluctuating fluid mechanical forces in which the time dependence is derived from blades passing other blades, struts, and so on. Another class of vibrations resulting from *instabilities* can arise when fluid does work on a vibrating blade (or other part) to amplify or maintain the vibration. Blade flutter is such a vibration. It is a very complex subject and one of great current research interest, so it will not be discussed in any comprehensive way here. Only a physical description of the mechanisms of blade flutter will be given.

Consider first the circumstances in which an isolated blade may flutter in bending. In the cascade representation of this blade, bending results in a vibrating velocity $\dot{z}$ more or less perpendicular to its chord, as shown in figure 7.11. As it vibrates the direction of the incident flow is changed as indicated by the dashed vector. From the curve of blade force F_z versus incidence, at the right of figure 7.11, we see that if the blade is operating in the *unstalled* range of incidence, say at point A, the force F_z is reduced by the motion $\dot{z}$, and this will retard or damp the motion. On the other hand, if the blade is operating in the *stalled* range of incidence, F_z is increased by $\dot{z}$, increasing the force in the direction that will cause the motion to grow. Thus we conclude that flutter is possible whenever the blade is at stalling incidence.

This argument suggests that unstalled blades will not flutter. This is true for isolated blades but not for cascades, as may be seen by examining the influence of neighboring blades of the cascade on any particular blade (reference 7.3). Consider the cascade sketched in figure 7.12, where the blade on which we wish to find the forces is denoted by 0, its first neighbor in the positive z direction by $+1$, and so on. Now in an infinite cascade, the blades must all flutter with the same amplitude because nothing distinguishes one from another, but there can be a phase shift from one to the next. In figure 7.12, the $+1$ and -1 blades are drawn dashed as they would appear if the phase shift from blade 0 to blade $+1$ is $\pi/2$. That is, when blade 0 has maximum positive velocity $\dot{z}$ and zero displacement, blade $+1$ will have maximum positive displacement, and blade -1 will have maximum negative displacement. The sketch shows that the streamtube between blades -1 and $+1$ is then wider than for the nonvibrating cascade. If the turning across the cascade is not decreased much by the vibration, the widened streamtube will result in an increased force on blade 0 in the direction of its motion; hence, the motion will be amplified. We conclude that an unstalled cascade can flutter in bending for an interblade phase angle $\psi = \pi/2$.

Now suppose the force perturbation ΔF_z can be represented as the sum of ΔF_{zip}, which would result if all blades had uniform spacing (if they vibrated in phase), and $\Delta F_{z\psi}$, which is due to the phase shift between blades. Both force perturbations should be proportional to dynamic pressure $\rho V^2/2$, so

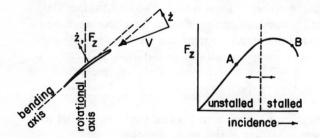

7.11   Force diagram for isolated blade vibrating in bending mode, showing possibility of flutter in stalled range of incidence.

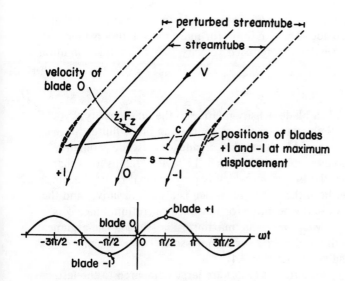

7.12   Diagram of compressor cascade in bending flutter.

we put $\Delta F_z = \Delta c_z (\rho V^2 / 2)(ch)$, where c is the blade chord and h is the blade height. By the above physical argument,

$$\Delta c_{zip} = m_{ip} \dot{z}_0 / V \tag{7.22}$$

and

$$\Delta c_{z\psi} = m_\psi (z_{+1} - z_{-1})/2s. \tag{7.23}$$

m_{ip} and m_ψ are coefficients of proportionality, and from figure 7.11, $m_{ip} < 0$ for unstalled incidence and $m_{ip} > 0$ for stalled incidence. $\dot{z}_0$ is the velocity of the zeroth blade. Let $z_0 = a \sin \omega t$; then $z_{+1} = a \sin(\omega t + \psi)$ and $z_{-1} = a \sin(\omega t - \psi)$, where ψ is the interblade phase angle. Substituting these in (7.23), we find $\Delta c_{z\psi} = m_\psi (a/s) \cos \omega t \sin \psi$.

Flutter will occur provided the work done on the blade 0 by the fluid in one complete vibrational period is large enough to balance the mechanical damping work. The fluid work ΔW_a may be written

$$\Delta W_a = \tfrac{1}{2}\rho V^2 (ch) \int_0^{2\pi/\omega} (\Delta c_{zip} + \Delta c_{z\psi})\dot{z}_0 \, dt;$$

this yields

$$\Delta W_a = \tfrac{1}{2}\rho V^2 (ch)\pi a [(m_{ip}/V)a + m_\psi (a/s) \sin \psi].$$

If we define a mechanical logarithmic decrement δ_m as the fractional decrease in vibrational energy per cycle due to mechanical damping, then the damping work per cycle ΔW_m is $\Delta W_m = -\delta_m m_b(\omega^2 a^2/2)$ where m_b is the effective mass of the blade. The condition for marginal stability is then $\Delta W_a + \Delta W_m = 0$. If we define a reduced velocity $\overline{V} \equiv V/c\omega$, this condition becomes

$$m_\psi \sin \psi \overline{V}^2 + (m_{ip} s/c)\overline{V} - \tfrac{1}{4}(s/c)\mu\delta_m = 0 \qquad (7.24)$$

where $\mu \equiv 4m_b/\pi c^2\rho h$ is a blade density relative to the fluid density.

This result exhibits the main parameters controlling blade flutter. Both m_ψ and m_{ip} are functions of Mach number, incidence, cascade geometry, and sometimes Reynolds number. Given their behavior (7.24) shows that flutter will occur when the first term, which is proportional to $\overline{V}^2$, becomes large enough to offset both the last term, which is always negative, and the second term, which is usually negative for unstalled operation. Since ψ is arbitrary, it can assume whatever value maximizes $|m_\psi \sin \psi|$. The argument above suggests this value is $\psi = \pi/2$ for compressors. This is roughly confirmed by detailed analysis (reference 7.4).

Ordinarily, the first two terms of (7.24) are large compared to the last, so the limiting value of $\overline{V}$ is almost independent of fluid density, which only appears in the last term.

References

7.1 S. Timoshenko, *Vibration Problems in Engineering*, Van Nostrand, 1937.

7.2 F. O. Carta, "Coupled Blade-Disc-Shroud Flutter Instabilities in Turbine Engine Rotors," *Journal of Engineering for Power*, July 1967, p. 419.

7.3 S. I. Ginsburg, "Calculation of the Boundary of Excitation of Flexural Flutter of a Homogeneous Array in a Quasistationary Approximation," *Strength of Materials* (Plenum Press), No. 8 (August 1974), pp. 46–51.

7.4 D. S. Whitehead, "Bending Flutter of Unstalled Cascade Blades at Finite Deflection," ARC R & M No. 3386, Her Majesty's Stationery Office, 1965.

Problems

7.1 Show that according to the discussion of section 7.1.1, the burst speed for a disc (without blades) is given by

$$\omega^2 = (\sigma_{\text{ultimate}}/\rho) \int_{r_I}^{r_H} z(r)\, dr \bigg/ \int_{r_I}^{r_H} r^2 z(r)\, dr$$

where r_I and r_H are the inner and outer radii and $z(r)$ is the axial thickness. Generalize this expression to include the effect of blades mounted on the outer radius of the disc.

7.2 A popular form of construction for modern aircraft engine compressors is the "drum rotor," in which the blades of all stages are mounted on a thin-walled drum having a radius equal to the hub radius of the blading. A disc with inner rim such as that shown in figure 7.2 supports the drum from the inside, carrying the centrifugal load of the blades. Following the method of section 7.1.1, develop a procedure for design of such a structure for constant stress.

7.3 In film cooling, the cooling air is passed through small holes on the skin of the blade. If the cooling air is colder than the blade skin, it will cool the interior surface of these small holes, generating a tensile stress. Estimate this stress by modeling the cooling hole as a long hole in a block of metal whose temperature is uniform far from the hole. Use (6.10) to estimate the heat flux at the wall of the hole.

7.4 A turbine mounted between two bearings as in figure 7.5 has a total mass, shaft plus discs plus blades, of 100 kg. It is operating at 10,000 rpm, which is three-fourths of its shaft critical speed, with a tangential velocity of 500 m s^{-1}, when one of 50 blades, with a mass of 0.5 kg, detaches from the rim. Estimate first the bearing force that results for very stiff bearings and second the shaft deflection at the bearings for very soft bearing mounts with $n_b/\omega_n = 0.5$.

8
Component Matching and Engine Performance

This chapter briefly explores how the performance and behavior of an actual engine is related to the characteristics of its components. Our view is now different from that taken in chapters 2 and 3 where the components were characterized by thermodynamic parameters π, τ, η with no reference to the shape or size of the machine. Here we begin with the characteristics of actual components, expressed by performance maps for the compressor (figure 5.22), turbine (figure 6.15), inlet (figure 4.25), fan (figure 5.20), and perhaps other components if necessary. These maps represent the behavior of real devices whose geometry is fully specified. Our task is to determine how they interact when combined into an engine or propulsion system.

The task is most conveniently divided into two parts. The first is termed component matching and involves applying the constraints that result from the need for the components to work together; the second involves predicting the performance of the resulting assembly.

8.1 Compressor-Turbine Matching: The Gas Generator

Suppose that we have a compressor, with the map of figure 5.22, and a turbine with the map of figure 6.15. With a combustor, the two mounted on a single shaft form a *gas generator*, which is the heart of any gas turbine engine. With an inlet and nozzle, the gas generator becomes a turbojet (figure 1.4). With a fan and fan drive turbine, it becomes a turbofan (figure 1.6). With a power turbine, it becomes a shaft turbine (figure 1.7), or it can be used as a direct source of hot, high-pressure air for innumerable other applications.

Matching of the compressor, combustor, and turbine implies for the station numbers of figure 1.4:

$$N_t = N_c \quad \text{or} \quad N_c/\sqrt{\theta_2} = (N_t/\sqrt{\theta_4})(T_{t4}/T_{t2})^{1/2} \tag{8.1}$$

$$W_4 = (1 + f)W_2 \quad \text{or} \quad \frac{W_4\sqrt{\theta_4}}{\delta_4 A_4} = (1 + f)\left(\frac{W_2\sqrt{\theta_2}}{\delta_2 A_2}\right)\left(\frac{p_{t2}}{p_{t4}}\right)\left(\frac{T_{t4}}{T_{t2}}\right)^{1/2}\frac{A_2}{A_4} \tag{8.2}$$

$$W_2 c_{pc}(T_{t3} - T_{t2}) = W_4 c_{pt}(T_{t4} - T_{t5}) \quad \text{or}$$

$$1 - (T_{t5}/T_{t4}) = \left(\frac{c_{pc}}{(1 + f)c_{pt}}\right)\left(\frac{T_{t2}}{T_{t4}}\right)\left(\frac{T_{t3}}{T_{t2}} - 1\right). \tag{8.3}$$

The connection between f, T_{t4}, and T_{t3} is given by

$$\frac{hf}{\bar{c}_p T_{t2}} = \frac{T_{t4}}{T_{t2}} - \frac{T_{t3}}{T_{t2}}. \tag{8.4}$$

If we specify two independent variables, usually $N_c/\sqrt{\theta_2}$ and T_{t4}/T_{t2}, then with these relations, the compressor and turbine maps, and a pressure drop relation such as (4.26) for the burner, we can determine the mass flow $W_2\sqrt{\theta_2}/\delta_2$, pressure ratio p_{t5}/p_{t2}, temperature ratio T_{t5}/T_{t2}, and fuel flow $hf/c_p T_{t2}$ for the gas generator.

In general the process is involved and tedious, but it is straightforward when the turbine nozzles are choked. Because this is the usual situation at full power, we will deal only with it. If the turbine nozzles are choked, then $W_4\sqrt{\theta_4}/A_4\delta_4$ has a unique value, determined by the geometry of the turbine nozzles, and (8.2) becomes an explicit expression for p_{t3}/p_{t2} as a function of $W_2\sqrt{\theta_2}/A_2\delta_2$ and T_{t4}/T_{t2}:

$$\frac{p_{t3}}{p_{t2}} = \left[\frac{(1+f)(A_2/A_4)}{\pi_b(W_4\sqrt{\theta_4}/A_4\delta_4)} \right] \frac{W_2\sqrt{\theta_2}}{A_2\delta_2} \left(\frac{T_{t4}}{T_{t2}} \right)^{1/2}. \tag{8.5}$$

The quantities in brackets are constant in first approximation. This is the equation of a straight line through the origin on the compressor map, as shown in figure 5.22. Its slope depends on the values of T_{t4}/T_{t2} and A_4/A_2;

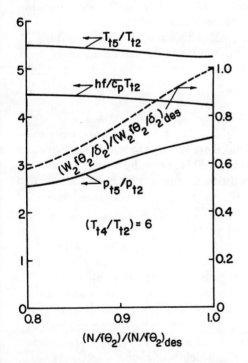

8.1 Pumping characteristics for a gas generator based on the compressor of figure 5.22, with $T_{t4}/T_{t2} = 6$.

thus if we select a design value of T_{t4}/T_{t2}, the turbine nozzle area must be selected relative to compressor area A_2 to put the operating line in the most advantageous portion of the compressor map.

Once the turbine geometry has been set, (8.5) gives p_{t3}/p_{t2} as a function of $N/\sqrt{\theta_2}$ and T_{t4}/T_{t2}. For a given $N/\sqrt{\theta_2}$, raising T_{t4}/T_{t2} increases p_{t3}/p_{t2} and in excess this will stall the compressor.

Now if we specify T_{t4}/T_{t2} and $N/\sqrt{\theta_2}$, the operating point on the compressor map is determined; hence p_{t3}/p_{t2}, η_c, and therefore T_{t3}/T_{t2} are determined. From (8.4), $hf/\bar{c}_p T_{t2}$ is determined, but f is calculable only if T_{t2} is specified. If we neglect f compared to unity, T_{t5}/T_{t4} from (8.3) depends only on $N/\sqrt{\theta_2}$ and T_{t4}/T_{t2}. Finally, η_t is determined by $N/\sqrt{\theta_4}$ and T_{t5}/T_{t4}, so p_{t5}/p_{t4} can be found. Thus except for the determination of f, which requires T_{t2}, we can find $W_2\sqrt{\theta_2}/\delta_2$, p_{t5}/p_{t2}, T_{t5}/T_{t2}, and $hf/\bar{c}_p T_{t2}$ as functions of $N/\sqrt{\theta_2}$ and T_{t4}/T_{t2}. These are called the "pumping characteristics" of the gas generator.

The pumping characteristics for a gas generator using the compressor of figure 5.22, with $T_{t4}/T_{t2} = 6$ and a turbine efficiency $\eta_t = 0.90$, are shown in figure 8.1. We see that at 100 percent corrected speed the gas generator produces a stagnation pressure ratio of 3.30 with a stagnation temperature ratio of 5.20. A set of curves such as those in figure 8.1 can be found for each value of T_{t4}/T_{t2}.

8.2 Gas Generator-Nozzle Matching

Two independent parameters, $N/\sqrt{\theta_2}$ and T_{t4}/T_{t2}, are required to specify the operating point of the gas generator. But if we specify the size of the exit nozzle A_n, this is reduced to one parameter. This is seen by writing the statement that the nozzle mass flow equals the turbine mass flow or $W_n = (1 + f)W_2$, so

$$\frac{W_n\sqrt{\theta_n}}{A_n\delta_n} = (1 + f)\left(\frac{W_2\sqrt{\theta_2}}{A_2\delta_2}\right)\left(\frac{A_2}{A_n}\right)\left(\frac{p_{t2}}{p_{t5}}\right)\left(\frac{T_{t5}}{T_{t2}}\right)^{1/2} \tag{8.6}$$

If the nozzle is choked, $W_n\sqrt{\theta_n}/A_n\delta_n$ has a specific value. If it is not choked, $W_n\sqrt{\theta_n}/A_n\delta_n$ depends on p_{t5}/p_0. In either case we can regard (8.6) as determining the nozzle/compressor area ratio required for operation at any particular set of T_{t4}/T_{t2} and $N/\sqrt{\theta_2}$. Alternatively, if A_n/A_2 is fixed, (8.6) gives a relationship between T_{t4}/T_{t2} and $N/\sqrt{\theta_2}$, so we can find the corrected speed at which the fixed-nozzle engine will run as a function of T_{t4}/T_{t2}. The engine now has a single control variable, T_{t4}/T_{t2}, which is of course directly related to fuel flow rate.

8.3   Engine-Inlet Matching and Distortion

The mass flow of the engine must also match that of the inlet. Since the corrected speed $N/\sqrt{\theta_2}$ and hence the corrected weight flow $W_2\sqrt{\theta_2}/\delta_2$ of the compressor are determined by T_{t4}/T_{t2} for the fixed-nozzle engine, the inlet must provide a variable Mach number M_2 at the engine face in response to changes in T_{t4}/T_{t2} and it must do so over a range of flight Mach numbers M_0. This is the reason for the complex variable-geometry inlets discussed in section 4.2. They have variable throat area, provision for bypassing air around the engine, and so on, to allow the inlet to operate with good pressure recovery while meeting the corrected weight-flow requirement of the engine.

No discussion of inlet-engine matching would be complete without some reference to inlet distortion, one of the most troublesome problems in modern propulsion systems. The flow the engine receives from the inlet is not uniform either radially or circumferentially. Even in a perfectly axisymmetric inlet there would be regions of low stagnation pressure near the outer walls. If the inlet is at angle of attack or side-mounted on a fuselage, there may be large circumferential variations. Some inlets have S-bends (as the center inlets on the B-727 and L-1011) that produce strong secondary flows.

The effect of these nonuniformities is to lower the stall margin of the fan or compressor that first receives the flow. If it has sufficient tolerance to distortion, it will reduce the nonuniformities thus shielding the following components; if not, then its stall may result in a general breakdown of flow throughout the engine.

Because the response of the compressor to distortion depends on the details of the flow, which may be unsteady and have variations in stagnation pressure and temperature both radially and circumferentially, no simple description of the inlet-compressor interaction has been successful in accurately predicting when stall will occur. This can be determined only by running the engine with the inlet in place. Approximate techniques have been developed for use in the design process, however. They consist of estimating or measuring the flow expected from the inlet, then representing it as a combination of radial and circumferential distortion patterns such as are sketched in figure 8.2. The patterns are characterized by the radial or circumferential extent of the low p_{t2} region, and its amplitude in the form

$$\frac{(p_{t2})_{\max} - (p_{t2})_{\min}}{(p_{t2})_{\max}}.$$

Distortion patterns approximating these are generated in the engine test facility by inserting screens of the required radial or circumferential extent

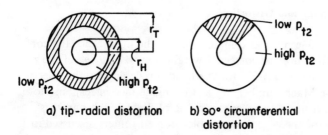

a) tip-radial distortion   b) 90° circumferential
                               distortion

8.2   Distortion patterns used for evaluating compressor response to inlet flow distortion.

and the required density in the flow upstream of the engine. The stall line is then determined with distortion.

Such steady-state distortion testing is now routine for all new aircraft propulsion systems.

Unfortunately, it is not always sufficient because the inlet outflow may be quite unsteady, so that *instantaneously* the distortion may be worse than the mean represented by the steady-state test. Extensive studies have shown that a compressor will stall if a distortion pattern that would lead to steady-state stall persists for a time of the order of the flow time through the compressor or longer (references 8.1, 8.2). Much stronger distortions can be tolerated for very short time periods.

8.4   Overall Performance

The pumping characteristics exemplified by figure 8.1 make it relatively simple to estimate the thrust and specific impulse (or the power output and specific fuel consumption). Consider the turbojet for simplicity. The thrust is given by (1.12), which is best written

$$\frac{F}{A_n p_0} = \left(\frac{W_2\sqrt{\theta_2}}{A_2\delta_2}\right)\left(\frac{\delta_2 u_0}{\sqrt{\theta_2}p_0 g}\right)\left(\frac{A_2}{A_n}\right)\left[(1 + f)\frac{u_e}{u_0} - 1\right] + \frac{A_e}{A_n}\left(\frac{p_e}{p_0} - 1\right). \qquad (8.7)$$

From the gas generator characteristics, we have $W_2\sqrt{\theta_2}/A_2\delta_2$, p_{t5}/p_{t2}, and T_{t5}/T_{t2}. The area ratios A_n/A_2 and A_e/A_n must be set. Then u_e/u_0 follows from

$$\frac{u_e}{u_0} = \frac{M_e}{M_0}\left(\frac{\gamma_t R_t T_e}{\gamma_c R_c T_0}\right)^{1/2} = \frac{M_e}{M_0}\left(\frac{\gamma_t R_t T_{te}}{\gamma_c R_c T_{t0}}\right)^{1/2}\left(\frac{1 + \frac{1}{2}(\gamma_c - 1)M_0^2}{1 + \frac{1}{2}(\gamma_t - 1)M_e^2}\right)^{1/2}; \qquad (8.8)$$

for the nonafterburning turbojet $T_{te}/T_{t0} = T_{t5}/T_{t2}$ and for the afterburning turbojet $T_{te}/T_{t0} = T_{t6}/T_{t2}$ (see figure 1.4). The exit Mach number is determined by the nozzle pressure ratio p_{t6}/p_0 and area ratio A_7/A_n. For a choked convergent nozzle $M_e = 1$. As A_7/A_n is increased from unity, $M_e = M_7$

increases as shown by figure 4.1, but a γ lower than 1.4 should be used for good accuracy. Having found M_e, one can determine p_e/p_0 by

$$p_e/p_0 = (p_{t6}/p_{t2})(\pi_d\delta_0)/[1 + \tfrac{1}{2}(\gamma_t - 1)M_e^2]^{\gamma_t/(\gamma_t - 1)} \tag{8.9}$$

and the expression for F/A_np_0 is complete.

The specific impulse I may be written

$$I \equiv \frac{F}{\dot{m}_f g} = \left[\frac{(F/A_np_0)}{(W_2\sqrt{\theta_2}/A_2\delta_2)(hf/\bar{c}_pT_{t2})}\right]\left(\frac{A_np_0}{A_2\delta_2}\right)\left(\frac{h}{\bar{c}_pT_{t2}}\right)(\sqrt{\theta_2}). \tag{8.10}$$

Two interesting observations can be made from (8.7) and (8.10). First, from (8.7), for a given flight Mach number M_0, a given θ_2 and a given value of T_{t4}/T_{t2}, the thrust is directly proportional to the ambient pressure. T_0 is nearly constant in the stratosphere so this proportionality will apply there. Second, from (8.10), we see that I is independent of p_0 but varies as $\sqrt{\theta_2/T_{t2}}$ $\propto 1/\sqrt{T_{t2}}$, so the specific impulse increases as T_0 decreases at fixed M_0.

Once the performance of an engine is established by test at sea level static conditions, its performance can be estimated by these means over a wide range of Mach numbers, engine speeds, and altitudes. The estimated thrust, specific fuel consumption, air flow, and bypass ratio for the JT3D-1 (see figure 1.16) are shown in figure 8.3 for an altitude of 35,000 ft.

8.5 Control and Acceleration

Gas turbines require sophisticated control systems because they generally operate at speeds and temperatures close to their limits of durability. Especially in aircraft engines, the range of environmental conditions is large and so is the number of variables that the control system must deal with. For an afterburning turbofan engine we might list the following *control variables*:

primary fuel flow rate

afterburner fuel flow rate

exhaust nozzle area

guide vane and stator angles

bleed valve settings

To control these variables, the control system might sense some or all of the following *measurables*:

inlet temperature T_{t2}

inlet pressure p_{t2}

compressor discharge pressure p_{t3}

turbine blade temperature T_{tr}

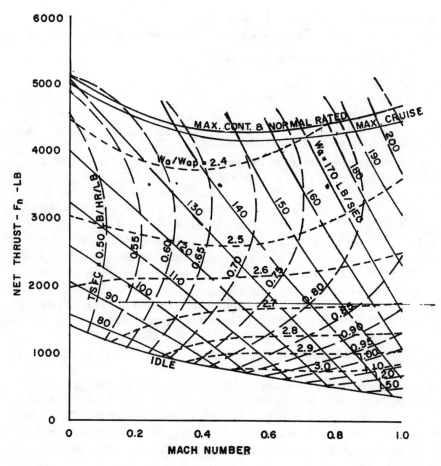

8.3   Estimated thrust, specific fuel consumption and airflow of JT3D engine (see figure 1.16) at 35,000 ft altitude.

turbine outlet temperature T_{t5}

turbine outlet pressure p_{t5}

gas generator speed N_2

fan speed N_1

The pilot interacts with the engine by setting the position of a power lever, which in turn makes an input to the control system, but the pilot does not directly control any of the above control variables. Instead, the power lever position selects a thrust level from idle to maximum, and the control manipulates the control variables to give the desired thrust level while observing the operating limits of the engine.

The limiting parameters differ from engine to engine, but physically they stem from speed limits set by stresses in the rotating parts, turbine temperature limits, compressor or fan pressure ratio limits set by stall or surge, and blade Mach number limits set by flutter or vibration. The stress, blade temperature, presence of stall or flutter are not sensed directly by the control; they are inferred from the measurements listed above except in some very recent engines, where the turbine bucket temperature is actually monitored. The control must therefore have a built-in logic that allows it to schedule the variables in accordance with the sensed measurables to observe the actual physical limits.

Further, the control must operate in the hot, vibrating environment of the engine nacelle with high reliability. Until very recently, this has dictated that the controls be hydromechanical, that is, that all the logic be performed with levers, cams, and flowing fuel or air. This has resulted in complex and expensive mechanisms and some real limitations on the functions of the control. For example, in current engines the motion of all compressor stators is ganged together as a function of $N/\sqrt{\theta_2}$, whereas better performance could be had by varying the several rows individually.

Recently advances in integrated circuits have made a digital electronic control feasible, and it will doubtless be incorporated in all sophisticated engines built in the 1980s. This will permit a more complex logical system to be used and should result in better performance as well as improved reliability.

The details of controls are too diverse to permit any description here. Good descriptions are given in references 8.3 and 8.4. The remainder of this discussion will deal with the engine acceleration dynamics.

In addressing the engine dynamics, we must be aware of the various time scales that characterize events in the engine. In order of increasing time they are air flow times, of the order of engine length/flow speed $\sim 10^{-2}$ s; fuel flow times $\sim 10^{-1}$ s; guide vane and nozzle response times, limited by control power, $\sim 10^{-1}$ s; and angular acceleration time, limited by engine power and surge margin, ~ 1–10 s.

8.5.1 Acceleration

Because of the long times required for angular acceleration of the engine, we can assume in treating this acceleration process that the components of the engine behave as they would in steady state except for the power matching between the turbine and other components. Consider a turbojet for simplicity; the turbine power is not equal to the compressor power during acceleration, since part of the turbine power goes to increasing the rotational energy of the rotor. We can write the turbine torque as

$$\mathcal{T}_t = \frac{\text{turbine power}}{\text{angular velocity}} = \frac{c_p T_{t4}(1 - \tau_t)W_2}{\omega};$$

the compressor torque is

$$\mathcal{T}_c = \frac{c_p T_{t2}(\tau_c - 1)W_2}{\omega}.$$

Thus if J is the angular moment of inertia of the rotor, the angular acceleration $d\omega/dt$ is given by

$$J(d\omega/dt) = \mathcal{T}_t - \mathcal{T}_c = [(c_p T_{t2}W_2)/\omega][(T_{t4}/T_{t2})(1 - \tau_t) - (\tau_c - 1)].$$

In terms of corrected parameters

$$d(N/\sqrt{\theta_2})/dt = \left[\frac{c_p T_0 \delta_2 (W_2 \sqrt{\theta_2}/\delta_2)}{4\pi^2 J \sqrt{\theta_2}(N/\sqrt{\theta_2})}\right][(T_{t4}/T_{t2})(1 - \tau_t) - (\tau_c - 1)]. \tag{8.11}$$

Now for steady-state operation, the quantity $f = (T_{t4}/T_{t2})(1 - \tau_t) - (\tau_c - 1)$ is zero. From section 8.1, it can be regarded as a function of $N/\sqrt{\theta_2}$, T_{t4}/T_{t2}, and A_n/A_2; thus we can represent f by an expansion around the steady-state operating condition

$$f = \frac{\partial f}{\partial(N/\sqrt{\theta_2})}\left[\frac{N}{\sqrt{\theta_2}} - \left(\frac{N}{\sqrt{\theta_2}}\right)_0\right] + \frac{\partial f}{\partial(T_{t4}/T_{t2})}\left[\frac{T_{t4}}{T_{t2}} - \left(\frac{T_{t4}}{T_{t2}}\right)_0\right]$$

$$+ \frac{\partial f}{\partial(A_n/A_2)}\left[\frac{A_n}{A_2} - \left(\frac{A_n}{A_2}\right)_0\right],$$

where the subscript zero denotes the value of each of the variables at the steady state. Suppose that the engine is running at one steady-state condition and step changes are made in A_n/A_2 or T_{t4}/T_{t2}, which are then held constant while $N/\sqrt{\theta_2}$ changes to its new steady-state value. During the transient in $N/\sqrt{\theta_2}$, A_n/A_2, and T_{t4}/T_{t2} both have their new steady-state values, so the equation for $N/\sqrt{\theta_2}$ becomes

$$\frac{d}{dt}\left[\frac{N}{\sqrt{\theta}} - \frac{N}{\sqrt{\theta_2}}\right)\right] = \frac{c_p T_0 \delta_2 (W_2 \sqrt{\theta_2}/\delta_2)}{4\pi^2 J \sqrt{\theta_2}(N/\sqrt{\theta_2})_0}\left[\frac{\partial f}{\partial(N/\sqrt{\theta_2})}\right]_0\left[\frac{N}{\sqrt{\theta_2}} - \left(\frac{N}{\sqrt{\theta_2}}\right)_0\right]$$

$$= \frac{-1}{t_{acc}}\left[\frac{N}{\sqrt{\theta_2}} - \left(\frac{N}{\sqrt{\theta_2}}\right)_0\right]$$

where

$$t_{acc} \equiv -\frac{4\pi^2 J \sqrt{\theta_2}(N/\sqrt{\theta_2})_0}{c_p T_0 \delta_2 (W_2 \sqrt{\theta_2}/\delta_2)_0}\left[\frac{1}{\partial f/\partial(N/\sqrt{\theta_2})}\right]_0$$

is a characteristic time for acceleration. The solution is

$$(N/\sqrt{\theta_2}) - (N/\sqrt{\theta_2})_0 = \text{const } e^{-t/t_{acc}}$$

so the corrected speed relaxes exponentially from its original value as shown in figure 8.4.

The time t_{acc} required for the engine to accelerate is independent of whether the change in $(N/\sqrt{\theta_2})_0$ is due to a change in T_{t4}/T_{t2} or in A_n/A_2.

To obtain the magnitude of t_{acc} we must estimate $\partial f/\partial(N/\sqrt{\theta_2})$. We can do this approximately as follows. We use an asterisk to denote each of the quantities in f divided by its value at the steady-state; thus

$$(N/\sqrt{\theta_2})/(N/\sqrt{\theta_2})_0 = (N/\sqrt{\theta_2})^*,$$

$$\tau_c/(\tau_c)_0 = \tau_c^*;$$

and we assume that

$$(W_2\sqrt{\theta_2}/\delta_2)^* = (N/\sqrt{\theta_2})^*,\tag{8.12}$$

$$(W_2\sqrt{\theta_4}/\delta_4)^* = 1 \quad \text{(choked turbine nozzle)},\tag{8.13}$$

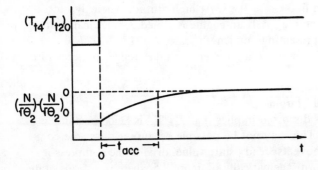

8.4   Response of the corrected speed to a step change in its steady-state value.

$(W_2\sqrt{\theta_5}/A_n\delta_5)^* = 1$ (choked exit nozzle),   (8.14)

$\eta_t = \eta_c = 1$.   (8.15)

From (8.13) $(W_2\sqrt{\theta_2}/\delta_2)^*(p_{t2}/p_{t4})^* = 1$; so using (8.15) and (8.12) gives

$(\tau_c^*)^{\gamma/(\gamma-1)} = \pi_c^* \approx (p_{t4}/p_{t2})^* = (N/\sqrt{\theta_2})^*\sqrt{(T_{t4}/T_{t2})^*}$.

From (8.14), $(N/\sqrt{\theta_2})^*((T_{t5}/T_{t4})^{1/2})^*(1/\pi_c^*\pi_t^*) = 1$, and using (8.15) gives

$(\tau_t^*)^{-(\gamma+1)/2(\gamma-1)} = (\tau_c^*)^{\gamma/(\gamma-1)}/(N/\sqrt{\theta_2})^* = 1$.

Now

$f = (T_{t4}/T_{t2})_0(T_{t4}/T_{t2})^*[1 - (\tau_t)_0\tau_t^*] - [(\tau_c)_0\tau_c^* - 1]$

$= (T_{t4}/T_{t2})_0[1 - (\tau_t)_0] - (\tau_c)_0(N/\sqrt{\theta_2})^{*(\gamma-1)/\gamma} + 1$,

and

$$\left[\frac{\partial f}{\partial(N/\sqrt{\theta_2})}\right]_0 = \frac{1}{(N/\sqrt{\theta_2})_0}\left[\frac{\partial f}{\partial(N/\sqrt{\theta_2})^*}\right]_0 = -(\tau_c)_0\left(\frac{\gamma-1}{\gamma}\right)\frac{1}{(N/\sqrt{\theta_2})_0}.$$ (8.16)

Thus the acceleration time becomes

$$t_{acc} = \frac{4\pi^2 J\sqrt{\theta_2}(N/\sqrt{\theta_2})^2}{c_p T_0\delta_2(W_2\sqrt{\theta_2}/\delta_2)_0}\frac{[\gamma/(\gamma-1)]}{(\tau_c)_0}$$

$$= \frac{4\pi^2 JN^2}{c_p T_{t2}W_2}\left(\frac{\gamma/(\gamma-1)}{(\tau_c)_0}\right).$$ (8.17)

This result has a relatively simple interpretation, namely that the numerator represents the rotational energy of the rotor $J\omega^2$, while the denominator represents the available accelerating power, which is clearly proportional to $c_p T_{t2}W_2$.

For a typical turbojet with $W_2 = 50$ kg s^{-1}, $N \approx 200$ s^{-1}, $\tau_c \approx 2$, $T_{t2} = 290°$K, $J \approx 5$ kg m^2, we find $t_{acc} \approx 0.8$ s. At high altitude where $W_2 \propto p_0$ is reduced, it can be much longer. If an engine is scaled geometrically, at constant tip speed, L representing any length scale, $J \propto L^5$, $N \propto L^{-1}$, $W_2 \propto L^2$, and we find $t_{acc} \propto L$, so a large engine takes longer to accelerate than a small one.

8.5.2 Acceleration Stall Margin

The acceleration process discussed implies that T_{t4}/T_{t2} is held constant during the acceleration. This may not be possible because it raises the compressor pressure ratio above its steady-state value, driving the compressor toward stall. This is shown schematically in figure 8.5 where the steady-state operating line is shown; $N/\sqrt{\theta_2}$ increases with T_{t4}/T_{t2} as indicated by the

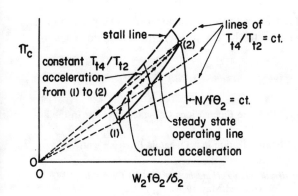

8.5   Showing the steady-state operating line for a fixed nozzle engine, and the transient followed in a constant T_{t4}/T_{t2} acceleration, driving compressor into stall.

points. If the engine is operating at point 1, and T_{t4}/T_{t2} is suddenly raised to the value required for steady operation at 2, π_c rises along the constant $N/\sqrt{\theta_2}$ line as shown and then follows the new T_{t4}/T_{t2} line as the engine accelerates. The situation shown would lead to stall so it is unacceptable; the control system would restrict the actual increase in T_{t4}/T_{t2} to prevent stall, but then the acceleration time would be greater than t_{acc} as given by (8.17). Nevertheless, some excess compression ratio is always required for acceleration. It is called the acceleration stall margin.

8.5.3   Other Transients
When the afterburner is lit on a turbojet engine (figure 1.4), the nozzle throat area must be increased simultaneously; otherwise the increased backpressure on the turbine will cause the engine to decelerate. But the timing is not very critical because of the long time scale for change of $N/\sqrt{\theta_2}$. In an afterburning turbofan, on the other hand, this timing is critical, because the back-pressure from an afterburner light is communicated within milliseconds to the fan (figure 2.7) and may cause it to stall. If it does, the compressor may follow and a general breakdown in the flow will result. This problem has existed in all military afterburning turbofans to date. The solution lies in a complex control that minimizes the timing error between afterburner light and nozzle opening and in afterburner fuel systems that give a nearly continuous variation of T_{t6} without sudden steps.

Sudden transients may also result from hot gas ingestion into the compressor, when the compressor suddenly is shifted to a lower corrected speed, while the pressure in the combustor is still that corresponding to the higher corrected speed. Stall is a very probable result if the hot gas persists for a compressor flow time or more. This situation occurs in military aircraft when

the ordnance (guns or rockets) are fired, and some aircraft have provision
for momentarily reducing fuel flow to the engines when the guns or rockets
are fired.

References

8.1 G. A. Plourde and B. Brimelow, "Pressure Fluctuations Cause Compressor Instability,"
AFAPL–TR–69–103, June 1970.

8.2 B. Brimelow and T. P. Collins, "Engine Testing in a Dynamic Environment," AIAA
Paper 74–1198, October 1974.

8.3 Pratt & Whitney Aircraft, *The Aircraft Gas Turbine Engine and Its Operation*, PWA
Operating Instruction 200, June 1952 (revised May 1974).

8.4 I. E. Treager, *Aircraft Gas Turbine Engine Technology*, McGraw-Hill, 1970.

Problems

8.1 Carry out the compressor-turbine matching described in section 8.1 to
find all of the pumping characteristics for one value of $(N/\sqrt{\theta_2})/(N/\sqrt{\theta_2})_{des}$
on figure 8.1.

8.2 Repeat the calculation leading to figure 8.1 for several values of T_{t4}/T_{t2},
compute the nozzle area A_n/A_2 required to operate the engine at design speed
at $T_{t4}/T_{t2} = 6$, and finally determine the variation of $(N/\sqrt{\theta_2})/(N/\sqrt{\theta_2})_{des}$
with T_{t4}/T_{t2} for this fixed A_n/A_2.

8.3 Using the pumping characteristics of figure 8.1, calculate the thrust and
specific impulse for a turbojet with convergent nozzle and a mass flow of
100 kg s^{-1} at sea level static conditions, taking $T_{t4}/T_{t2} = 6$ at the design
point. Assume $\pi_d \approx 0.95$.

8.4 When a very rapid increase in thrust is desirable, as for example in a
wave-off from an aircraft carrier, and a variable-area nozzle is available,
it might be desirable to keep the (turbojet) engine at full rpm, lowering the
thrust for approach by opening the nozzle. By suddenly closing the nozzle,
thrust can be recovered without the delay associated with angular acceleration
of the engine. Describe what happens in the engine when the nozzle is suddenly closed.

8.5 The acceleration dynamics of a two-shaft turbofan differs from that of a
turbojet because the core accelerates independently of, and much faster than,
the fan and its associated turbine. Develop a model analogous to that of
section 8.5.1 for a turbofan. Assume that the core changes speed instantly.

8.6 In a duct-burning turbofan with a fan pressure ratio of 3, the afterburner is inadvertently lit with the nozzle still in the nonafterburning position, raising T_{t6} by a factor of two. Compute the resulting change in fan outlet pressure. If the same thing occurs in a turbojet, what will be the result?

9
Aircraft Engine Noise

Though long of concern to neighbors of major airports, aircraft noise first became a major problem with the introduction of turbojet-powered commercial aircraft (B-707, Dehavilland Comet) in the late 1950s. These aircraft were powered by engines originally developed for military aircraft. It was recognized at the time that the noise levels produced by military aircraft would be unacceptable to persons living under the takeoff pattern of major airports such as New York's Kennedy (then Idlewild) and London's Heathrow. Accordingly, much effort was devoted to developing jet noise suppressors (reference 9.1), with some modest success. Takeoff noise restrictions were imposed by some airport managements, notably the Port of New York Authority, and nearly all first-generation turbojet-powered transports were equipped with jet noise suppressors at a significant cost in weight, thrust, and fuel consumption.

The introduction of the turbofan engine with its lower jet velocity (Rolls-Royce Conway, Pratt & Whitney JT3-D) temporarily alleviated the jet noise problem but increased the high-frequency turbomachinery noise, which became a severe problem on landing approach as well as takeoff. This noise was reduced somewhat by proper choice of rotor and stator blade numbers and spacing and by using engines of single mixed-jet type.

In spite of these efforts, the increasing volume of air traffic resulted in unacceptable noise exposures near major urban airfields in the late 1960s, leading to great public pressure for noise control. This pressure stimulated and advancing technology enabled the making of Federal Aviation Rule, Part 36 (FAR-36), which set maximum takeoff, landing, and "sideline" noise levels for certification of new turbofan-powered aircraft. The B-747, DC-10, and L-1011 meet this rule.

Figure 9.1 shows the three measuring stations referred to in FAR-36: under the approach path one mile before touchdown, under the takeoff path 3.5 miles from the start of the takeoff roll, and along the sides of the runway

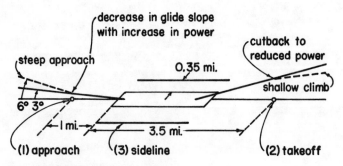

9.1 Schematic of airport runway showing approach 1, takeoff 2, and sideline 3 noise measurement stations.

at a distance of 0.35 mile (for four-engined aircraft). The noise at the approach and takeoff stations depends on both the engines and the aircraft performance and operational procedures, since the power settings and aircraft height can be variables. The sideline station is more representative of the intrinsic takeoff noise characteristics of the engine, since the engine is at full throttle and the station is at a fixed distance from the aircraft.

Because of this rule airport noise is now a major design criterion for civil aircraft, so an understanding of the mechanisms of noise production and of the techniques for alleviating it is crucial for aircraft propulsion research and development.

There are substantial differences between the noise characteristics of subsonic and supersonic engines. As indicated schematically in figure 9.2, the subsonic turbofan radiates noise forward and backward from its large, high tip speed fan. It also produces jet noise from both the fan jet and the primary jet, but because a low jet velocity gives good propulsive efficiency in cruise, the jet noise can be reduced while improving fuel economy and range. Thus for the subsonic high-bypass turbofan, fan noise is the most critical problem, both on approach and on takeoff.

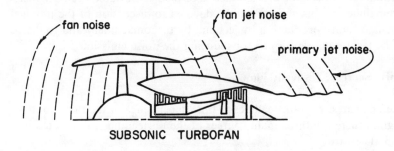

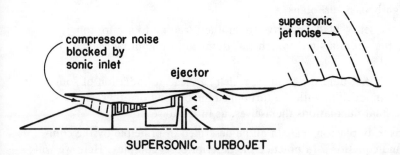

9.2 Schematic illustration of noise sources from subsonic turbofan engines (top) and supersonic turbojet engines (bottom).

The supersonic turbojet engine has a long inlet that can be operated choked on approach to suppress compressor noise, so that approach noise is not critical. If optimized for cruise performance, it has a high jet velocity on takeoff. The SST also has a high thrust/weight ratio on takeoff (on the order of 0.32 as compared to 0.25 for subsonic aircraft) so that after liftoff it can climb very rapidly. This reduces the noise at the takeoff measuring point, particularly if the engines are throttled back just before that point. But the full noise impact of the high thrust and jet velocity is sensed at the sideline station.

Supersonic transports were not covered by FAR-36, and neither the present generation of SST's (Concorde, Tu 144) nor the original design for the US SST (B-2707) are or were capable of meeting the requirement on sideline noise. This was one of the factors that led to cancellation of the B-2707 development.

It now appears that to provide satisfactory subsonic and supersonic performance and also meet a noise rule equivalent to or more stringent than FAR-36, an advanced SST engine will have to be capable of operation with a high-bypass ratio on takeoff, converting to a lower bypass for transonic acceleration and supersonic cruise. Such is termed a "variable cycle engine."

The remainder of this chapter will be devoted to discussion of the physical phenomena of noise production by jets and by turbomachinery and to discussion of the techniques developed for reducing the noise emission.

9.1 Noise Sources: Unsteady Flow

All noise emanates from unsteadiness—time dependence in the flow. In aircraft engines there are three main sources of the unsteadiness, illustrated in figure 9.3. They are

motion of the blading relative to the observer, which if supersonic can give rise to propagation of a sequence of weak shocks, leading to the "buzz saw" noise of high-bypass turbofans;

motion of one set of blades relative to another, leading to a pure-tone sound (like that from a siren) which was dominant on approach in early turbojets; and

turbulence or other fluid instabilities, which can lead to radiation of sound either due to interaction with the turbomachine blading or other surfaces, or from the fluid fluctuations themselves, as in jet noise.

These unsteady phenomena can all be described in principle by the compressible fluid equations. In practice the description is complex. Here we will attempt only a qualitative treatment sufficient for conveying a physical understanding of the phenomena.

TURBOMACHINE NOISE

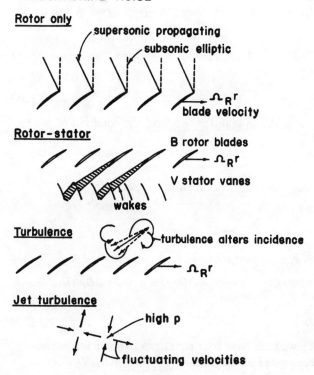

9.3   Sources of unsteady flow giving rise to noise from aircraft engines.

Since most readers of this book are expected to have a classical background in fluid mechanics, we begin with the equations of conservation of mass

$$D\rho/Dt + \rho\nabla\cdot\mathbf{u} = 0 \qquad (9.1)$$

and conservation of momentum (inviscid)

$$\rho D\mathbf{u}/Dt = -\nabla p \qquad (9.2)$$

where $D/Dt \equiv \partial/\partial t + \mathbf{u}\cdot\nabla$ is the "substantial derivative" with respect to time, following the fluid motion. If the fluid is assumed non-heat-conducting as well as inviscid, then the entropy is constant for a fluid element as it is carried along by the flow, that is,

$$Ds/Dt = 0. \qquad (9.3)$$

To distinguish sound or noise from other small disturbances in the flow, we first linearize these equations, taking $\mathbf{u} = \mathbf{u}_0 + \mathbf{u}'$, $p = p_0 + p'$, and so on, where $\mathbf{u}_0$ and p_0 are uniform in space and time. Dropping terms quadratic in the small disturbances, we find

$$D\rho'/Dt + \rho_0 \nabla \cdot \mathbf{u}' = 0, \tag{9.1a}$$

$$\rho_0 D\mathbf{u}'/Dt = -\nabla p', \tag{9.2a}$$

$$Ds'/Dt = 0, \tag{9.3a}$$

where now $D/Dt = \partial/\partial t + \mathbf{u}_0 \cdot \nabla$. In addition, we have the equation of state $p = \rho RT$, which yields

$$p'/p_0 = \rho'/\rho_0 + T'/T_0, \tag{9.4}$$

and the connection between entropy, pressure, and temperature

$$s - s_0 = c_p \ln T/T_0 - R \ln p/p_0,$$

which when "linearized" is

$$s' = c_p T'/T_0 - Rp'/p_0 = c_v p'/p_0 - c_p \rho'/\rho_0.$$

Differentiating this gives $Ds'/Dt = 0 = (c_v/p_0)Dp'/Dt - (c_p/\rho_0)D\rho'/Dt$, so that

$$D\rho'/Dt = (\rho_0/\gamma p_0)(Dp'/Dt). \tag{9.5}$$

Substituting this in (9.1a), we then have four equations to solve for p' and $\mathbf{u}'$. The velocity can be eliminated by subtracting $\nabla \cdot$ (9.2a) from D/Dt (9.1a) to get

$$\frac{\rho_0}{\gamma p_0} \frac{D^2 p'}{Dt^2} - \nabla^2 p' = 0. \tag{9.6}$$

This is a *wave equation* for p', with the wave velocity $a_0 = (\gamma p_0/\rho_0)^{1/2}$, so we identify this quantity as the sound velocity.

Our basic relation for what follows is (9.6), but if we take $\nabla \times$ (9.2a), then since $\nabla \times \nabla \equiv 0$,

$$\frac{D}{Dt}(\nabla \times \mathbf{u}') = 0. \tag{9.7}$$

Since $\nabla \times \mathbf{u}' = \omega'$ is the vorticity disturbance, this says that small amplitude vorticity is convected and has associated with it no pressure fluctuations. Equations (9.3a), (9.$\bar{6}$), and (9.7) describe three independent disturbance modes: entropy, sound, and vorticity.

Now consider the behavior of sound as described by (9.6). For simplicity, we take $\mathbf{u}_0 = 0$. No generality is lost because we can transform to a coordinate system moving at $\mathbf{u}_0$ without changing the wave behavior. The equation is then

$$\frac{\partial^2 p'}{\partial t^2} - a_0^2 \nabla^2 p' = 0.$$

9.1.1 Waves, Acoustic Power, Decibels

For one-dimensional or plane waves we have $\partial^2 p'/\partial t^2 - a_0^2 \partial^2 p'/\partial x^2 = 0$, which is satisfied by any function having the argument $x \pm a_0 t$, that is, $p' = p'(x \pm a_0 t)$. This simply says that p' is constant along "characteristics" $x = \pm a_0 t$. To determine the actual form and magnitude of p' we must specify some boundary conditions on the solution. Suppose, for example, we want the sound radiated by a plane perpendicular to the x-axis, vibrating according to $x = A \sin \omega t$, so that its velocity is $\dot{x} = A\omega \cos \omega t$. To find the velocity in the wave we return to (9.2a) which is, for this case, $\rho_0(\partial u'/\partial t) = -\partial p'/\partial x$. Now if we assume a solution $p = P \cos k(x - a_0 t)$, then

$$\frac{\partial u'}{\partial t} = -\frac{1}{\rho_0}\frac{\partial p'}{\partial x} = +\frac{Pk}{\rho_0}\sin k(x - a_0 t),$$

and integrating gives $u' = (P/\rho_0 a_0) \cos k(x - a_0 t)$. Matching this to the boundary condition requires $P/\rho_0 a_0 = A\omega$ and $-ka_0 = \omega$.

Now the energy transferred across a plane by this wave per cycle per unit area is the force per unit area times velocity integrated over a cycle, or

$$\int_0^{2\pi/\omega} p'u'\,dt = \int_0^{2\pi/ka_0} (P^2/\rho_0 a_0)\cos^2 k(x - a_0 t)\,dt = \left(\frac{P^2}{\rho_0 a_0}\right)\left(\frac{\pi}{ka_0}\right). \tag{9.8}$$

The power per unit area, $d\mathbf{P}/dA$ is $ka_0/2\pi$ times this, so $d\mathbf{P}/dA = P^2/2\rho_0 a_0$, where $P^2/2$ will be recognized as the mean square pressure fluctuation.

For a spherical wave emanating from a point source, for example, (9.6) is

$$\frac{\partial^2 p'}{\partial t^2} - a_0^2 \frac{1}{r^2}\frac{\partial}{\partial r}\left[r^2 \frac{\partial p'}{\partial r}\right] = 0,$$

and we find

$$p' = (Pr_0/r)\cos k(r - a_0 t),$$
$$u' = (P/\rho_0 a_0)(r_0/r)\cos k(r - a_0 t) - (P/\rho_0 a_0)(r_0/kr^2)\sin k(r - a_0 t), \tag{9.9}$$

while the sound power per unit area is $d\mathbf{P}/dA = (P^2/2\rho_0 a_0)(r_0/r)^2$. It dies off as $1/r^2$, but the total power radiated over spherical surfaces surrounding the radiator is constant.

Because the human ear's response to sound is such that the perceived sound is proportional to the logarithm of the pressure fluctuation, both the sound pressure level and sound power level are quoted in decibels dB, where

sound pressure level, dB $= 20 \log_{10} (p'/.00002)$, where p' is in Nm^{-2}

and

sound power level, dB $= 10 \log_{10} (P/10^{-12})$, where P is in watts.

For reference, a pressure of $1\ Nm^{-2} = 94\ dB$, while one atmosphere $(1.015 \times 10^5\ Nm^{-2}) = 194.1\ dB$.

9.1.2 Monopoles, Dipoles, Quadrupoles

We can think of p' given by (9.9) as resulting from the harmonic expansion and contraction of a small sphere of radius r_0, the amplitude of pressure fluctuation at its surface being P. This vibrating sphere might model the acoustical effect of a pulsating jet, as indicated in figure 9.4. In the limit as $r_0 \to 0$ with Pr_0 a constant, (9.9) is the solution for a point source, or monopole, with acoustic power

$$P_m = \text{monopole power} = 2\pi(Pr_0)^2/\rho_0 a_0. \tag{9.10}$$

It is a fundamental solution of (9.6) from which other solutions can be constructed by superposition.

Two of special importance here are the dipole and the quadrupole.

The dipole is composed of two monopoles of opposite signs, or shifted in phase by π if harmonic, separated by some distance d as shown in figure 9.4. Because they are equal in magnitude but opposite in sign, the mass flows of the two just cancel, but as they oscillate fluid is accelerated back and forth between the two, leading to an *oscillating force* that can represent the effect of a vibrating blade, as sketched in figure 9.4. To find the sound radiated by such a dipole, we add two monopole solutions (9.9). The essential feature is that if the measurement point is at a distance r from one monopole, it is at a slightly different distance, approximately $r + d \cos \theta$, from the other; so the sound waves arriving at the measurement point from the two monopoles must have left them at times differing by roughly $d \cos \theta/a_0$. Thus,

$$p' = (Pr_0/r) \cos k(r - a_0 t) - [Pr_0/(r + d \cos \theta)] \cos k\left[r - a_0\left(t - \frac{d \cos \theta}{a_0}\right)\right].$$

Expanding the second term, we find

$$p' \approx (Pr_0/r)(kd \cos \theta) \sin k(r - a_0 t)$$

if $kr \gg 1$, so that the measurement point is many wavelengths from the dipole. The radiated power from the dipole is $4\pi r^2(p')^2/\rho_0 a_0$, or

$$P = \text{Dipole power} = \frac{2\pi(Pr_0)^2}{\rho_0 a_0} (kd \cos \theta)^2. \tag{9.11}$$

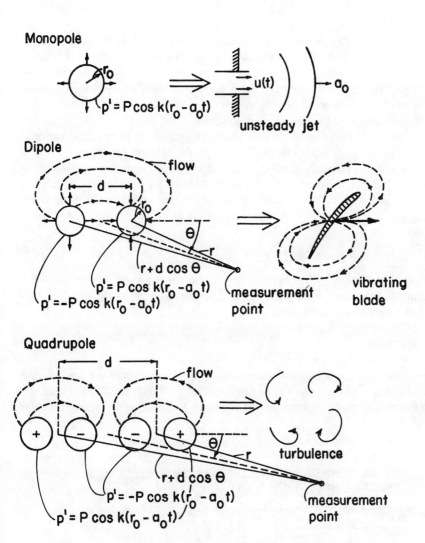

Monopole

$p' = P \cos k(r_0 - a_0 t)$

unsteady jet

Dipole

flow

$p' = P \cos k(r_0 - a_0 t)$

$p' = -P \cos k(r_0 - a_0 t)$

$r + d \cos \theta$

measurement point

vibrating blade

Quadrupole

flow

turbulence

$p' = -P \cos k(r_0 - a_0 t)$

$p' = P \cos k(r_0 - a_0 t)$

$r + d \cos \theta$

measurement point

9.4   Monopole, dipole, and quadrupole sources and some fluid disturbance they can represent.

Comparing to (9.10), we see that the dipole power is $(kd \cos \theta)^2$ times that of the monopole. The directional characteristic is embodied in $\cos^2 \theta$. The factor $kd = 2\pi d/\lambda$ is the ratio of spacing between the monopoles to the wavelength λ of the radiated sound, so for a given spacing d set by the geometry of the blading which the dipole represents, the acoustical power of the dipole goes down as the frequency is lowered.

If we combine two dipoles so that the force they exert on the fluid cancels, we have the simplest representation of a fluid disturbance involving *no mass injection and no externally applied force*, but only fluid pressures and accelerations which are in balance according to (9.2). Such a disturbance can still radiate sound, as we see by repeating this argument, separating two dipoles by a distance d to form the (linear) quadrupole sketched in figure 9.4. Then we find

$$\mathbf{P}_q = \text{Quadrupole power} = [2\pi(Pr_0)^2/\rho_0 a_0](kd)^4; \tag{9.12}$$

the directional behavior has been neglected. So for the same level of pressure fluctuation, the quadrupole radiates $(kd)^2$ as much power as the dipole, and $(kd)^4$ as much as the monopole.

9.2 Jet Noise

When fluid issues as a jet into a stagnant or more slowly moving background fluid, the shear between the moving and stationary fluids results in a fluid mechanical instability that causes the interface to break up in violent turbulent

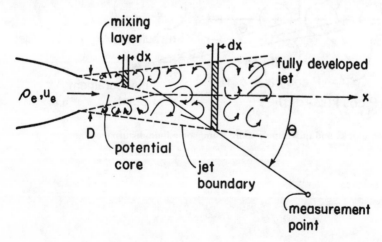

9.5 Subsonic jet mixing with ambient air, showing mixing layer followed by fully developed jet.

fluctuations as indicated in figure 9.5. The characteristics of the instability
are different for supersonic jets than for subsonic ones.

9.2.1 Subsonic Jets

For the subsonic jets, Lighthill (reference 9.2) has shown that the turbulence in
the jet can be viewed as a distribution of quadrupoles. The argument is complex, but physically it is equivalent to the following. The scale of the turbulent
fluctuations and hence of the quadrupoles is D, the jet diameter, and the
magnitude of the pressure fluctuations is $\rho_e u_e^2$, so $(Pr_0)^2$ becomes $(\rho_e u_e^2 D)^2$.
The wave number $k = \omega/a_0$, and $\omega \approx u_e/D$, so $kd \approx u_e/a_0$. Subsituting these
in (9.12) and taking $\rho_e \approx \rho_0$, we find

$$\mathbf{P}_j = \text{jet acoustic power} \approx 2\pi\rho_0 u_e^8 D^2/a_0^5; \tag{9.13}$$

since the argument is dimensional not quantitative, the factor of 2π is not to
be regarded seriously. The striking feature of this relation is the dependence
on u_e^8, which was first derived by Lighthill (reference 9.2). It has been verified
as correct for subsonic jets over a wide range of velocities.

It is instructive to rewrite the expression as

$$\mathbf{P}_j = \text{jet acoustic power} \approx D^2(\rho_e u_e^3)\frac{\rho_e}{\rho_0}(u_e/a_0)^5, \tag{9.14}$$

from which we identify $D^2(\rho_e u_e^3)$ as the jet kinetic power, and $(u_e/a_0)^5$ as the
fifth power of jet Mach number referred to the speed of sound in the ambient
air. Thus we can write the jet's "acoustic efficiency," $\eta_{\text{jet noise}}$, as

$$\eta_{\text{jet noise}} \propto (\rho_e/\rho_0)M_{e0}^5. \tag{9.15}$$

This expression will be modified later in comparison to jet noise data.

An important feature of jet noise is its directional characteristics. The
intensity is largest in directions about 45 degrees from the direction of the
jet, as shown in figure 9.6. This is explained (reference 9.3) as a result of two
effects. First, the quadrupoles are not stationary, but are convected along
by the flow at a Mach number which is about half M_{e0}. This tends to focus
their radiation downstream. The second effect is refraction. As the sound
propagates out of the jet into the still air, it is turned away from the axis,
leaving a quiet zone directly behind the jet.

The frequency distribution of the sound from a jet can be explained by
thinking of the jet as divided axially into a series of slices of thickness dx
(reference 9.3; see figure 9.5), each emitting sound at a frequency $\omega \approx u/d$
where u is the local jet velocity and d the local height of the mixing region.
The initial portion of the jet comprises an inviscid core and a linearly growing mixing region; here we take $u = u_e/2$, the volume element $dV \approx xD\,dx$,

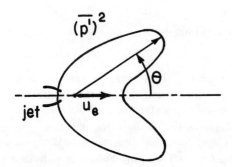

9.6   Directional characteristics of the noise from a subsnoic jet.

and $\omega \approx u/d = u_e/2x$. Now from (9.13), the acoustic power per unit of jet volume is about $\rho_0 u^8/La_0^5$ where L is interpreted as the distance along the jet to the radiating element, so the contribution of the elements in the mixing region, where $L \approx x$, is

$$d\mathbf{P} = (\rho_0 u_e^8 D/a_0^5)\, dx, \quad \text{and for these elements } \omega = u_e/2x.$$

In the fully developed part of the jet, u decreases so that $ux \approx u_e D$, and $\omega \approx u/d \approx u_e D/x^2$, while the volume element is $x^2\, dx$, and again $L \approx x$, so we have

$$d\mathbf{P} \approx \frac{\rho_0 u_e^8 D}{a_0^5}\left(\frac{D}{x}\right)^7 dx, \quad \text{and for these elements } \omega \approx u_e D/x^2\,.$$

We see that the mixing layer contributes to the high frequencies, the fully developed jet to the lower frequencies. To estimate the frequency spectrum we note that $d\mathbf{P}/d\omega = (d\mathbf{P}/dx)(dx/d\omega)$ and find

$$\frac{d\mathbf{P}}{d\omega} \approx \frac{\rho_0 u_e^9 D}{a_0^5}\,\omega^{-2}; \quad \text{(mixing region high frequencies)}$$

$$\frac{d\mathbf{P}}{d\omega} \approx \frac{\rho_0 u_e^5 D^5}{a_0^5}\,\omega^2; \quad \text{(developed jet low frequencies)}$$

$$(9.16)$$

Thus, $d\mathbf{P}/d\omega$ increases as ω^2 for low frequencies and decreases as ω^{-2} for high frequencies, as sketched in figure 9.7. The peak occurs for a frequency

$$(\omega/2\pi) \approx Sr(u_e/D) \tag{9.17}$$

where the Strouhal number Sr is between 0.15 and 0.20 for subsonic jets.

9.2.2   Supersonic Jets
This description must be modified for high-velocity jets where $M_{e0} > 2$. First of all, it is clear that the jet's acoustic efficiency as given by (9.15)

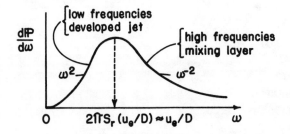

9.7   Frequency spectrum of jet noise, showing contributions of mixing layer and developed jet (from reference 9.3).

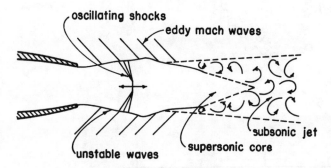

9.8   Schematic of supersonic jet, showing shock structure due to imperfect expansion, and unstable waves in supersonic mixing layer.

cannot exceed unity, so the M_{e0}^5 law must fail for large M_{e0}. In fact, experiments show that for large M_{e0}, the jet acoustic power varies as u_e^3, not u_e^8.

Two mechanisms exist for noise production that are not present in subsonic jets. They are shown schematically in figure 9.8. The supersonic shear layer is subject to a wave-like instability (reference 9.4) which leads to a slower growth than for the subsonic shear layer and to the radiation of "eddy Mach waves" when the rate of convection of the eddies is more than the speed of sound in the ambient air (when $M_{e0} \gtrsim 2$). If the jet is not perfectly expanded, then shocks occur in the adjustment to ambient pressure. Since these shocks are formed by the deflection of the jet boundary, they will be unsteady if the boundary is, and they can add considerably to the noise. This can add some 3 to 5 dB to the noise level for jets that are not perfectly expanded (reference 9.5). The effect is shown in figure 9.9, where data from a wide range of jet conditions are plotted. Here, to account for effects of jet Mach number, a modified jet acoustic efficiency has been used, $\mathbf{P}/\rho_0 A_e a_0^3 M_e^3$, where $M_e \neq M_{e0}$ but is rather $M_e = u_e/a_e$. It seems clear that M_e must govern the jet structure, not M_{e0}. The same factor is put in the abscissa,

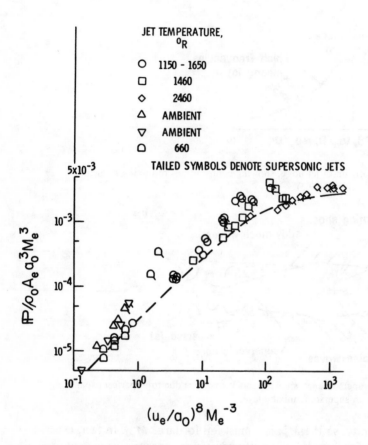

9.9 Jet acoustic power relative to jet kinetic power as a function of $(u_e/a_0)^8 M_e^{-3} = (u_e/a_0)^8(a_e/u_\epsilon)^3$ (from reference 9.5).

so that if $\mathbf{P}/\rho_0 A_e a_0^3 M_e^3$ were actually proportional to $(u_e/a_0)^8 M_e^{-3}$, the data would fall on a straight line, the Lighthill prediction. Some points fall above, due to shock noise (tailed symbols), while at large jet velocities, the data slopes off toward a constant value of about 0.003 for $\mathbf{P}/\rho_0 A_e a_0^3 M_e^3$, which is representative for rockets, and implies $\mathbf{P} \propto u_e^3$.

9.2.3 Jet Noise Suppression
Methods for suppressing jet noise have exploited the characteristics of the jet itself and those of the human observer. For a given total noise power, the human impact is less if the frequency is very high where the ear is less sensitive. Such a frequency shift can be achieved by replacing one large nozzle by many small ones, since $\omega \sim u_e/D$ where D is the jet diameter. This was one basis for the early turbojet engine suppressors (reference 9.1). Reduction

of the jet velocity can have a powerful effect ($P \propto u_e^8$ to u_e^3, depending on u_e). The multiple small nozzles gave some reduction in mean jet velocity by promoting entrainment of the surrounding air into the jet. Some attempts have been made to augment this effect by enclosing the multinozzle in a shroud, so that the ambient air is drawn into the shroud in the way described in section 4.3.2.

Certainly the most effective of jet noise suppressors has been the turbofan engine, which in effect distributes the power of the exhaust jet over a larger airflow, thus reducing the mean jet velocity.

In judging the overall usefulness of any jet noise reduction system, several factors must be considered in addition to the amount of noise reduction, namely the thrust loss, the added weight, and additional fuel consumption.

Consider the tradeoff between thrust and noise suppression. At takeoff the thrust F is proportional to $\dot{m}_e u_e$, while if $M_{e0} < 2$ (turbofans) we conclude from (9.15) that $P_j \propto \dot{m}_e u_e^7$. We can identify three interesting special cases:

1. *An ideal ejector*, in which the jet power is held constant as the mass flow is increased. Thus, $\dot{m}u_e^2$ is constant, and $F \propto (\dot{m}u_e^2)/u_e$ varies as $1/u_e$, while $P_j \propto (\dot{m}u_e^2)u_e^5$, so as u_e is lowered F increases and P_j decreases like u_e^5, or $P \propto F^{-5}$. The noise level in dB is $10 \log_{10} P$, so the noise reduction in decibels that results from changing from thrust F_0 to F by changing u_e is

$$\Delta dB = -50 \log_{10}(F/F_0); \quad \text{ideal ejector.} \tag{9.18}$$

This is plotted as $F/F_0 = 10^{-\Delta dB/50}$ in figure 9.10. Further, if $\dot{m}_0$ represents the mass flow at thrust F_0 (turbojet) then $\dot{m}/\dot{m}_0 = (u_{e0}/u_e)^2 = (F/F_0)^2 = 1 + \alpha$ where α is the bypass ratio. The resulting values of α are shown in figure 9.10, where we see that a bypass ratio of 2 gives a bit more than 10 dB reduction in jet noise. This was about the gain made in going from the JT-3 to the JT-3D (figure 1.16).

2. *Smaller engines* with the same velocity would give $P \propto \dot{m} \propto F$, so

$$\Delta dB = 10 \log_{10}(F/F_0) \tag{9.19}$$

which is plotted as the lowest curve on figure 9.10. Clearly, this is a bad solution.

3. *Constant mass flow*, with a reduction in u_e to effect the noise reduction, would give $P \propto F^7$ and

$$\Delta dB = 70 \log_{10}(F/F_0) \tag{9.20}$$

which is plotted as the middle full line in figure 9.10. Such a noise suppression could be obtained in a turbojet engine by operating at full corrected speed (full $W_2 \sqrt{\theta_2}/\delta_2$) with the exhaust nozzle opened as T_{t4}/T_{t2} is reduced to reduce u_e.

Measured against these sample cases, the results of actual noise suppres-

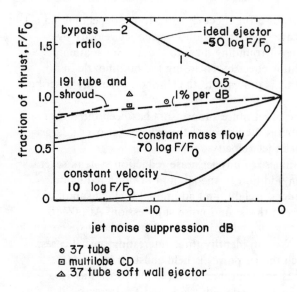

9.10 Variation of thrust with jet noise reduction, for ideal ejector (turbofan), constant mass flow and constant exhaust velocity engines, compared to experimental results.

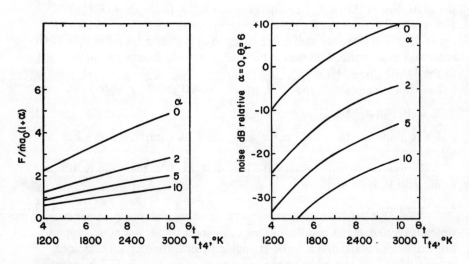

9.11 Variations of thrust and jet noise with bypass ratio α and turbine inlet temperature T_{t4} for turbofan engines.

sion programs (reference 9.6) are better than the constant-mass flow case, but far worse than the ideal ejector in terms of the penalty suffered to achieve a given level of noise suppression. Most suppressors cost about 1 percent loss in thrust per dB of noise reduction, including the very elaborate multitube ejectors studied for the U.S. SST.

The effects of both bypass ratio α and turbine inlet temperature T_{t4} on jet noise can be estimated readily from the simple cycle analysis of chapter 2. Figure 9.11 shows the variations of thrust per unit of total airflow and jet noise, the latter relative to $\theta_t = 6$, $\alpha = 0$, for the optimum turbofan discussed in section 2.5. Higher bypass ratios will be required to maintain acceptable jet noise levels as T_{t4} is increased.

9.3 Turbomachinery Noise

Turbomachinery generates noise by producing time-dependent pressure fluctuations, which can be thought of in first approximation as dipoles, since they result from fluctuations in force on the blades or from passage of lifting blades past the observer. As dipoles the blades radiate in accord with (9.11), and if we carry out the argument that led to (9.13) for jet noise, we find that for the blade noise

$$\mathbf{P}_{b\,lade} \approx \rho u^6 c^2/a^3 \approx \rho_2 w_2^3 c^2 M_T^3 \qquad (9.21)$$

where c is the blade chord or other characteristic dimension, and $u/a \approx M_T$, the tangential Mach number, leading to the second expression in the notation of chapter 5. By the argument of sections 1.9 and 5.1.2, the power exchanged with the fluid by the blade is of order $\rho_2 w_2^3 c^2 M_T^2$, so the acoustic efficiency

$$\eta_{\text{blade noise}} \propto M_T. \qquad (9.22)$$

However, the situation is more complex than this because in contrast to the exhaust jet, the rotating blade is not free to radiate sound to the observer; instead the sound waves must propagate out of the engine inlet or exhaust before they can be radiated. Because this propagation has a controlling influence on turbomachinery noise, it will be discussed in some detail.

9.3.1 Duct Modes
Suppose that a rotor or a rotor-stator pair is located at $z = 0$ in the annular duct of figure 9.12 and provides a source of acoustic excitation such that there is a pressure pattern

$$p'(r, \theta, 0) = g(r)e^{i(m\theta - \omega t)}. \qquad (9.23)$$

If the pattern is from a rotor with angular velocity Ω_R and m blades, then

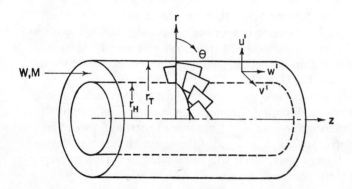

9.12   Schematic of a turbomachine duct with rotating blade row, illustrating notation
for duct propagation analysis.

$\omega = m\Omega_R$. We want to determine how this pressure pattern behaves in the
annular duct; in particular, we want to know whether it propagates or is
attenuated. We assume that there is a uniform, purely axial velocity W in the
duct with Mach number M.

The pressure field is governed by (9.6); when written in the cylindrical
coordinates appropriate to the duct geometry, it becomes

$$\frac{1}{a_0^2}\frac{\partial^2 p'}{\partial t^2} + \frac{2M}{a_0}\frac{\partial^2 p'}{\partial z \partial t} + (M^2 - 1)\frac{\partial^2 p'}{\partial z^2} - \frac{1}{r}\frac{\partial}{\partial r}\left(r\frac{\partial p'}{\partial r}\right) - \frac{1}{r^2}\frac{\partial^2 p}{\partial \theta^2} = 0.$$

If we suppose the solution is of the form $p' = f(r)e^{i(m\theta + kz - \omega t)}$, this is

$$\frac{1}{r}\frac{d}{dr}\left(r\frac{df}{dr}\right) + \left[-(1 - M^2)k^2 - \frac{2M\omega k}{a_0} + \left(\frac{\omega}{a_0}\right)^2 - \frac{m^2}{r^2}\right]f = 0$$

or, more compactly,

$$r^2\frac{d^2 f}{dr^2} + r\frac{df}{dr} + [\mu^2 r^2 - m^2]f = 0, \tag{9.24}$$

where $\mu^2 = -(1 - M^2)k^2 - 2M\omega k/a_0 + (\omega/a_0)^2$. This is a Bessel equation
with solutions

$$f(\mu r) = c_1 J_m(\mu r) + c_2 Y_m(\mu r), \tag{9.25}$$

where $J_m(\mu r)$ is the Bessel function of the first kind, of order m, and $Y_m(\mu r)$
is the Bessel function of the second kind. Both functions are tabulated and
plotted in reference 9.8. A plot of $J_{10}(\mu r)$ is given in figure 9.13. $Y_m(\mu r)$
behaves similarly for large μr, with a phase shift, but is infinite at $\mu r = 0$.

The solution for the present problem must satisfy the boundary condition
that the radial velocity perturbation be zero at the inner and outer walls,

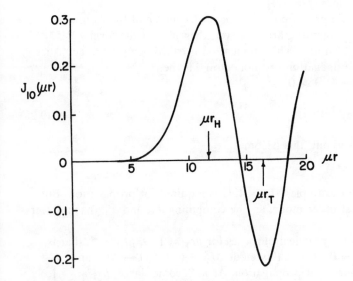

9.13   Bessel function of first kind of order 10.

$r = r_H, r_T$. Now from the radial component of (9.2a)

$$(\rho_0)(\partial u'/\partial t) = -(\partial p'/\partial r) \quad \text{or} \quad -i\rho_0\omega u(r) = -dp(r)/dr = -df/dr,$$

so the boundary condition is that $df/dr = 0$, $r = r_H, r_T$; with (9.25) we have two relations,

$$\begin{aligned} c_1 J_m'(\mu r_H) + c_2 Y_m'(\mu r_H) &= 0, \\ c_1 J_m'(\mu r_T) + c_2 Y_m'(\mu r_T) &= 0, \end{aligned} \tag{9.26}$$

which for a given value of r_H/r_T may be considered to determine μr_T and c_2/c_1. The solution can be carried out numerically, but to illustrate the results we can here regard r_H/r_T as a variable to be determined. Then since the point marked μr_T in figure 9.13 satisfies (9.26) for $c_2 = 0$, we will have a solution with just J_{10} if r_H/r_T is such that μr_H is the second point indicated. This gives $r_H/r_T = \mu r_H/\mu r_T = 11.9/16.5 = 0.72$, for $\mu r_T = 16.5$, $m = 10$.

Going back to the definition of μ, we can now write

$$(\mu r_T)^2 = -(1 - M^2)(kr_T)^2 - \frac{2M\omega r_T kr_T}{a_0} + \left(\frac{\omega r_T}{a_0}\right)^2$$

and since μr_T is known this becomes a quadratic relation for the axial wave number kr_T. If we further put $\omega r_T/a_0 = m\Omega_R r_T/a_0 = mM_T$, where M_T is the tangential Mach number of the blade tip

$$kr_T = \frac{-MM_T m \pm [m^2 M_T^2 - (\mu r_T)^2 (1 - M^2)]^{1/2}}{1 - M^2}. \tag{9.27}$$

Now if kr_T is real, the pressure disturbance is harmonic of the form e^{ik_Rz} in z; if kr_T has an imaginary part, however, then p' is of the form $e^{-k_Iz+ik_Rz}$ and hence is attenuated in z. This attenuation is called "cutoff" in acoustics. The condition for propagation or "noncutoff" is then $m^2 M_T^2 - (\mu r_T)^2(1 - M^2) > 0$, or

$$M_T^2 > \left(\frac{\mu r_T}{m}\right)^2 (1 - M^2). \tag{9.28}$$

For the mode worked out, this becomes

$$M_T^2 > (1.65)^2(1 - M^2).$$

Thus if $M = 0.5$, for example, $M_T > 1.43$ is required for propagation. But this is not the lowest order mode, as one combining J_{10} and Y_{10} has a lower $\mu r_T/m$.

A solution exists for practical purposes for $\mu r_T = 11.9$, $\mu r_H < 5$, that is, for $r_H/r_T < 5/11.9 = 0.42$. For this mode $M_T^2 > (1.19)^2(1 - M^2)$ and $M_T > 1.031$ is required for propagation. As m becomes large, $\mu r_T/m \to 1$, and the condition for propagation becomes simply $M_T^2 + M^2 > 1$; the relative Mach number to the blade tip must exceed unity for excitation of a

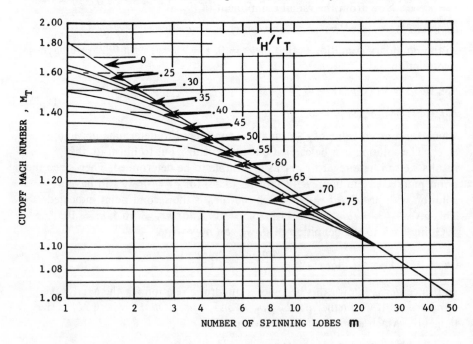

9.14 Cutoff tangential Mach number as a function of peripheral harmonic number m and hub/tip ratio r_H/r_T (from reference 9.9).

propagating mode. Cutoff tangential Mach number is shown as a function of r_H/r_T and m for $M = 0$ in figure 9.14.

If a mode is appreciably below cutoff, the imaginary part of kr_T, $k_I r_T$ is of order m, so the mode attenuates as $e^{-m(z/r_T)}$. For large m, then, the attenuation is rapid, and for practical purposes the cutoff modes do not radiate from the duct. It has been observed experimentally (reference 9.9) that the fan speed at which any given frequency first appears in the radiated sound agrees very well with (9.28).

9.3.2 Rotor-Stator Interaction

It would appear at first from this argument that compressors or fans should not radiate sound due to blade motion unless the blade tip speed is supersonic, but even low-speed turbomachines do in fact produce a great deal of noise at the blade passing frequencies. The explanation of this phenomenon was first presented by Tyler and Sofrin (reference 9.10). A simplified version is as follows. We consider a rotor with B blades in close proximity to a stator with V vanes, as indicated in figure 9.3. The stator blades will produce a lift force and hence a pressure field p' dependent on their incidence. Whatever the pressure field, it must be periodic with stator spacing, so let

$$p'_{\text{stator}} \propto (\text{incidence})e^{iV\theta}$$

so that when θ passes from 0 to 2π, p'_{stator} passes through V periods. The disturbance produced by the rotor is periodic with rotor spacing; in rotor coordinates,

$$(\text{incidence}) \propto e^{iB(\theta - \Omega_R t)}.$$

It follows that

$$p'_{\text{stator}} \propto e^{i[(V-B)\theta - B\Omega_R t]}.$$

Now this can be regarded as a pressure pattern rotating with angular velocity

$$\Omega_{\text{eff}} = \frac{d\theta}{dt} = \frac{B\Omega_R}{V - B} \tag{9.29}$$

which implies that the combination of rotor and stator will produce pressure patterns rotating faster than the rotor, by the factor $B/(V - B)$. This can be large if V and B are close together as they were in most early engines. In modern turbofans, $V > 2B$, so that rotor-stator interaction is no more likely to excite propagating modes than is the rotor rotation itself. No inlet guide vanes are used in large modern turbofans, in order to eliminate the interaction between their wakes and the rotor. A large axial gap between the rotor and its downstream stator decreases the interaction between them by allowing the rotor wakes to decay.

9.3.3 "Buzz Saw" or Combination Tone Noise

This analysis applied to an isolated rotor operating at an M_T above cutoff would predict a noise consisting of the blade-passing tone plus all of its higher harmonics (reference 9.11). No excitation of frequencies lower than blade-passing would be expected from a well-balanced rotor in which each blade is identical. What is in fact observed is a more or less random excitation of most of the harmonics of shaft rotational frequency up to and above the blade-passing frequency, as shown in figure 9.15. The large content of low frequencies is what leads to the characteristic "buzz saw" noise of high-bypass turbofans on takeoff.

There are at least two possible explanations for this fortunate occurrence—fortunate because if the turbofans did radiate at blade-passing, they would be very bad neighbors, as the blade-passing tone is near the peak of the ear's sensitivity. As indicated in figure 9.3, each blade near the tip generates a weak shock that propagates forward into the inlet annulus, but if the blades are not all identical, some of these shocks will be stronger than others and hence will propagate faster, overtaking their weaker upstream neighbors. When two shocks so interfere, they coalesce to form a stronger one, while a gap is left by the advancing shock. This mechanism can lead to the formation of a shock pattern that is a replica of the mechanical imperfections of the rotor

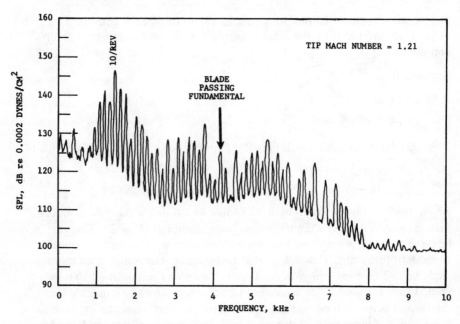

9.15 Harmonic analysis of "buzz saw" noise from large turbofan engine (from reference 9.9).

and hence would contain all harmonics of shaft rotational frequency (reference 9.12).

Another possible explanation lies in the behavior of the rotor blade wakes downstream of the rotor. It can be argued (reference 9.13) that the wake structure is unstable due to the strong mean swirl of the flow behind the rotor and that this instability leads to formation of disturbances downstream of the rotor with periodicity less than blade passing. The pressure field of these disturbances could influence the shock structure on the rotor, giving rise to the "buzz saw" tones.

9.3.4 Duct Linings

The treatment of propagation in section 9.3.1 assumed that the walls of the duct were rigid, so that the radial velocity u was zero there. One consequence of this is that no work can be done on the wall by the sound wave, so the wall takes no energy out of the wave. If the wall is modified so that pressure fluctuations at the wall result in a normal velocity component there, then energy can be extracted from the sound field, causing attenuation in the axial direction z.

This can be implemented in a number of ways; one is to line the duct with a porous sheet covering a series of small cavities, as shown in figure 9.16. Each small cavity functions as a "Helmholtz resonator"; the pressure in the cavity acts as a spring upon which the flow through the orifice oscillates in response to pressure fluctuations outside the orifice. Thus if p'_c is the pressure in the cavity and L is the effective length of the orifice,

$$\rho_0 L \frac{du'}{dt} = p' - p'_c - cu'.$$

c is a viscous drag coefficient and p'_c is connected to u' by

$$V(dp'_c/dt) = \rho_0 A u' = (V/a_0)(dp'_c/dt),$$

where A is the orifice area. Now, taking $p' = Pe^{i\omega t}$, we find

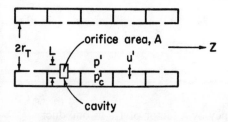

9.16 Schematic of acoustic damping cavities in an engine duct. The size of the resonators is exaggerated relative to the duct diameter.

$$\frac{d^2u'}{dt^2} + \frac{c}{\rho_0 L}\frac{du'}{dt} + \frac{Aa_0^2}{LV}u' = i\omega Pe^{i\omega t};$$

this has the solution $u' = Ue^{i\omega t}$ where

$$U = \frac{i\omega P/\rho_0 L}{\omega_n^2 - \omega^2 + i\omega c/\rho_0 L}.$$

Here $\omega_n^2 \equiv a_0^2 A/VL$ is the resonant frequency of the cavity. The ratio of u' to $p'/\rho_0 a_0$ is usually termed the "acoustical admittance"

$$\eta \equiv \frac{\rho_0 a_0 u'}{p'} = \frac{i\varepsilon\omega a_0/L}{\omega_n^2 - \omega^2 + i\omega c/\rho_0 L}, \tag{9.30}$$

where ε is the ratio of orifice area to wall area.

The effect on propagation is most easily seen by estimating the energy extracted from the duct disturbance by the resonators. The power extracted per unit of wall area is

$$\frac{dP}{dA_w} = \frac{\omega}{2\pi}\int_0^{2\pi/\omega} \text{Re}\,(p')\,\text{Re}\,(u')\,dt$$

$$= \frac{1}{2\pi}\int_0^{2\pi}\left(\frac{\varepsilon P^2 \cos\omega t}{\rho_0 L}\right)\frac{\omega[(\omega c/\rho_0 L)\cos\omega t - (\omega_n^2 - \omega^2)\sin\omega t]}{[(\omega_n^2 - \omega^2)^2 + (\omega c/\rho_0 L)^2]}$$

$$= (\varepsilon P^2\omega^2 c)/2(\rho_0 L)^2[(\omega_n^2 - \omega^2)^2 + (\omega c/\rho_0 L)^2].$$

The power in the wave propagating down the duct of diameter $2r_T$ is $P = \pi r_T^2 P^2/2\rho_0 a_0$ (see section 9.1), and $dA_w/dz = 2\pi r_T$, so

$$\frac{1}{P}\frac{dP}{dz} = -\frac{\omega^2 a_0 c\varepsilon}{\rho_0 r_T L^2[(\omega_n^2 - \omega^2)^2 + (\omega c/\rho_0 L)^2]}. \tag{9.31}$$

To estimate the attenuation length, suppose $cu' \approx \rho_0 u^2/2$, (all the dynamic head of flow through the orifices is lost), then $c \approx \rho_0 u'/2$ which we estimate as $\rho_0 a_0/2$. The decay length divided by the duct diameter is then

$$\frac{P}{2r_T}\frac{dz}{dP} = \frac{L^2\omega^2}{\varepsilon a_0^2}[(\omega_n^2/\omega^2 - 1)^2 + (a_0/2\omega L)^2].$$

Near resonance $\omega \approx \omega_n$ and this reduces to simply

$$\frac{P}{2r_T}\frac{dz}{dP} \approx \frac{1}{4\varepsilon} \tag{9.32}$$

so that to attenuate the resonant frequency by a factor of $1/e$ in one duct diameter requires $\varepsilon = 1/4$, that is, 25 percent open area in the duct lining.

The length scale of the openings is set by the frequency. We have $\omega_n^2 = a_0^2 A/VL$, and if we take $A \approx \varepsilon V/L$, $\omega_n^2 \approx a_0^2 \varepsilon/L^2$. Then for a rotor with B blades and angular velocity Ω_R, $L/r_T \approx \varepsilon^{1/2}/BM_T$. A honeycomb structure covered by a perforated plate is often used.

This analysis is meant only to illustrate the principles of duct attenuation. Detailed calculations and extensive experimentation are required to determine the attenuation properties of liners with engineering accuracy. A good summary of data is given in reference 9.14.

9.4 Noise Measurement and Rules

Human response sets the limits on aircraft engine noise, and while the logarithmic relationship represented by the scale of decibels is a first approximation to human perception of noise levels, it is not nearly quantitative enough for either systems optimization or regulation. Much effort has gone into the development of quantitative indices of noise. A review was given in reference 9.15. All that will be attempted here is a brief description of the major factors involved, and an explanation of the calculation procedures of Federal Aviation Rule, Part 36, which in a sense represents the official consensus on the noisiness of an individual aircraft.

9.4.1 Noise Effectiveness Forecast (NEF)
It is not the noise output of an aircraft per se that raises objections from the neighborhoods of major airports, but the total noise impact of the airport operations, which depends on takeoff patterns, frequencies of operation at different times of the day, population densities, and a host of less obvious things. There have been proposals to limit the total noise impact of airports, and in effect legal actions have done so for the most heavily used ones.

The accepted measure of noise impact at present is the Noise Effectiveness Forecast (NEF) which is arrived at as follows for any location near the airport.

1. For each event, compute the Effective Perceived Noise Level (EPNL) by the methods of FAR-36 to be described below. (This value is limited by FAR-36.)

2. For events occurring between 10 p.m. and 7 a.m., add 10 to the EPNdB.

3. Then NEF $= 10 \log_{10} \sum_i \log_{10}^{-1} (\text{EPNdB}/10)_i - 82$ where the sum is taken over all events in a 24-hour period.

A little ciphering will show that this last calculation is equivalent to adding for all the events the product of sound intensity times time, then taking the dB equivalent of this. The subtractor 82 is arbitrary.

As an example, suppose a point near the airport experiences 50 flyovers,

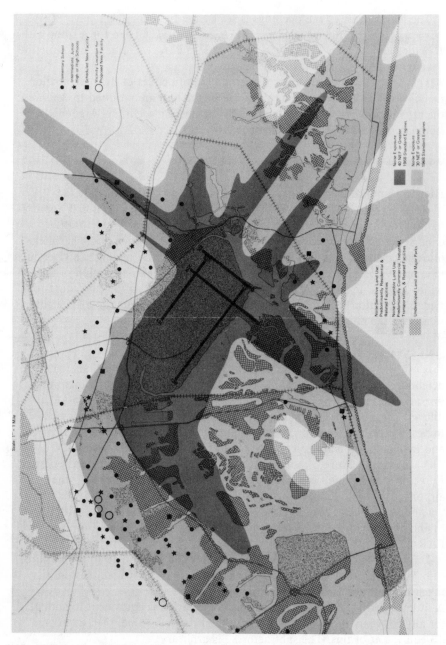

9.17   NEF = 40 and NEF = 30 contours for Kennedy Airport (1968), locations of schools, and land use patterns (from reference 9.15).

each imposing an EPNdB level of 100; then

$$NEF = 10 \log_{10}[50 \log_{10}^{-1} (100/10) - 82] = 35.$$

Major complaints occur when NEF exceeds about 30. As an indication of the magnitude of the problem, consider figure 9.17 for Kennedy Airport.

9.4.2 Effective Perceived Noise Level (EPNL)
The perceived noisiness of an aircraft flyover depends on the frequency content, relative to the ear's response, and on the duration. The perceived noisiness is measured in NOYs and is shown for random noise in figure 9.18. Note the great sensitivity in the range of frequencies from 2000 to 5000 Hz. Pure tones, frequencies with pressure levels much larger than that of the neighboring random noise in the sound spectrum, are judged to be more annoying than an equal sound pressure in random noise, so a "tone correction" is added to their perceived noise level. A "duration correction" represents the idea that the total noise impact depends on the integral of sound intensity over time for a given event.

The raw data for an EPNL calculation would consist of time histories of the sound pressure (SPL) for each of the one-third octave frequency bands

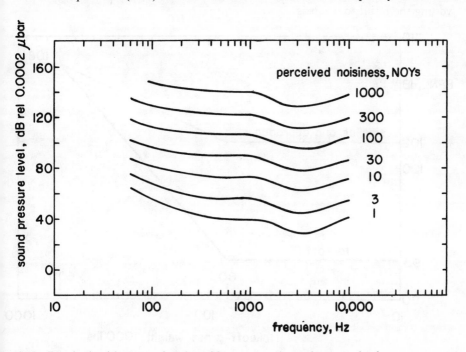

9.18 Perceived noisiness as a function of frequency and sound pressure level.

from 50 Hz to 10,000 Hz. Conceptually, the calculation of EPNL involves the following steps; for the exact prescription see reference 9.17.

1. Determine the NOY level for each band from figure 9.18 or equivalent, and sum them by the relation $N(k) = 0.85n(k) + 0.15 \sum_{i=1}^{24} n_i(k)$ where k denotes an interval in time, i the several frequency bands and $n(k)$ is the NOY level of the most noisy band. This reflects the "masking" of lesser bands by the noisiest.

2. The total PNL is then $\text{PNL}(k) = 40 + 33.3 \log_{10} N(k)$.

3. Apply a tone correction $c(k)$ by identifying the pure tones and adding to PNL an amount ranging from 0 to 6.6 dB, depending on the frequency of the tone and its amplitude compared to neighboring bands.

4. Apply a duration correction according to EPNL = PNLTM + D where PNLTM is the maximum PNL for any of the time intervals, and

$$D = 10 \log_{10}\left[\frac{1}{T}\sum_{k=0}^{d/t} \Delta t \log_{10}^{-1}\left[\text{PNLT}(k)/10 - \text{PNLTM}\right]\right]$$

where $\Delta t = 0.5$ sec, $T = 10$ sec, and d is the time over which PNLT exceeds PNLTM minus 10 dB. This amounts to integrating the sound pressure level over the time during which it exceeds its peak value minus 10 dB, then converting the result to decibels.

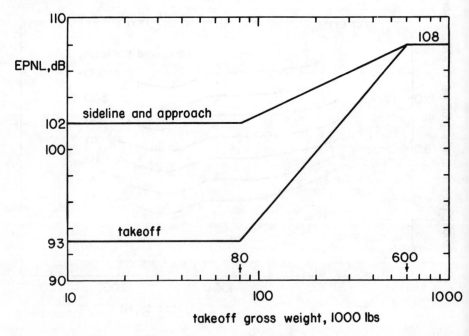

takeoff gross weight, 1000 lbs

9.19   Noise limits imposed by FAR–36 for certification of aircraft.

9.4.3 FAR-36 Noise Limits

Since December 1, 1969, all turbofan-powered aircraft with bypass ratios of 2 or more must comply *at certification* with EPNL limits for the three measuring stations shown in figure 9.1. The limits depend on takeoff gross weight of the aircraft as shown in figure 9.19. There is some flexibility, in that the noise levels shown can be exceeded by up to 2 EPNdB at any station provided the sum of the exceedances is not over 3 EPNdB, and the exceedances are completely offset by reductions at other measuring stations.

References

9.1 F. B. Greatrex and R. Bridge, "The Evolution of the Engine Noise Problem," *Aircraft Engineering*, Feb. 1967.

9.2 M. J. Lighthill, "On Sound Generated Aerodynamically, I: General Theory," *Proceedings of the Royal Society*, Series A, Vol. 211 (1952), pp. 564–587. Also: M. J. Lighthill, "Jet Noise," *AIAA Journal*, Vol. 1 (July 1963), pp. 1507–1517.

9.3 H. S. Ribner, "The Noise of Aircraft," in *Proceedings of the Fourth Congress of the International Council of Aeronautical Sciences*, Paris, 1964. Also: University of Toronto Institute of Aerospace Sciences, Report 24.

9.4 C. K. W. Tam, "Supersonic Jet Noise Generated by Large Scale Disturbances," AIAA Paper No. 73–992, Oct. 1973.

9.5 U. H. von Glahn, "Correlation of Total Sound Power and Peak Sideline OASPL from Jet Exhausts," AIAA Paper No. 72–643, June 1972.

9.6 G. S. Schairer, J. V. O'Keefe, and P. E. Johnson, "Perspective of the SST Aircraft Noise Problem," AIAA Paper No. 68–1023, Oct. 1968.

9.7 F. B. Hildebrand, *Advanced Calculus for Applications*, Prentice-Hall, 1962.

9.8 E. Jahnke and F. Emde, *Tables of Functions and Curves*, Dover, 1945.

9.9 M. J. Benzakein, "Research on Fan Noise Generation," *Journal of the Acoustical Society of America*, Vol. 51, No. 5, Part 1 (1972), pp. 1427–1438.

9.10 J. M. Tyler and T. G. Sofrin, "Axial Compressor Noise Studies," *SAE Transactions*, Vol. 70 (1962), pp. 309–332.

9.11 O. Okurounmu and J. E. McCune, "Three-Dimensional Vortex Theory of Axial Compressor Blade Rows at Subsonic and Transonic Speeds," *AIAA Journal*, Vol. 8 (1970), p. 1275.

9.12 M. R. Fink, "Shock Wave Behavior in Transonic Compressor Noise Generation," ASME Paper No. 71–GT–7, 1971.

9.13 W. T. Thompkins and J. L. Kerrebrock, "Exit Flow from a Transonic Compressor Rotor," MIT GTL Report No. 123, Sept. 1975. Also: *Unsteady Phenomena in Turbomachinery* (AGARD Conference Proceedings No. 177), 1975.

9.14 NASA Lewis Research Center, "Aircraft Engine Noise Reduction," NASA SP 311, May 1972.

9.15 J. W. Little and J. E. Mabry, "Human Reaction to Aircraft Engine Noise," AIAA Paper No. 68–548, June 1968.

9.16 National Academy of Sciences and National Academy of Engineering, *Jamaica Bay and Kennedy Airport: A Multidisciplinary Study*, Vol. 2, 1971.

9.17 "Aircraft Noise Evaluation," *Federal Register*, Vol. 34, No. 221, Appendix B, p. 18373.

Problems

9.1 Consider a wall with a shape given by $y = A \sin kx$ $(kA \ll 1)$, moving with a velocity U in the x direction in a compressible fluid with speed of sound a. Show that if $M \equiv U/a > 1$, sound waves are radiated to $y = +\infty$, while if $M < 1$, the pressure disturbance is attenuated in y. Compare these results to those of section 9.3.1.

9.2 Following the argument of section 9.1.2, work out the directional radiation characteristics of the linear quadrupole shown in figure 9.4, then repeat the argument for a rectangular or square quadrupole having sources of alternate sign on each side.

9.3 Taking the dipole model indicated in figure 9.4 for a vibrating blade, develop an approximate expression for the sound radiated by an airfoil of chord c oscillating with amplitude A in the direction perpendicular to its chord in an airflow of velocity U_0, pressure p_0, and temperature T_0.

9.4 A turbojet engine has a mass flow of 300 kg s^{-1} and a thrust of 300,000 N at takeoff. Estimate (a) its total radiated jet noise power, (b) the sound pressure level in dB at a distance of 0.35 mi, and (c) the frequency of peak sound intensity.

9.5 For the optimum turbofan of section 2.5, find the variations of thrust per unit of airflow and jet noise with bypass ratio, for $\theta_t = 6$, and check the results of figure 9.11.

9.6 A transonic fan without inlet guide vanes has a blade stagger angle (β_b') at the tip of $60°$, a hub/tip radius ratio of 0.4, and a tip radius of 1 m. Determine the rotative speed at which the rotor would be expected to first generate a propagating mode in the upstream duct at takeoff.

9.7 Consider two possible classes of subsonic transports, one of 600,000 lb gross weight, the other of 200,000 lb gross weight, but otherwise identical in aerodynamic performance, structural to gross weight ratio, engine performance, and so on, so that the smaller aircraft will carry one-third the number of passengers that the larger one does. For a given total number of passenger movements per day, how do the NEF levels for the two aircraft compare?

10
Hypersonic Engines

Airbreathing propulsion systems currently operate routinely at flight Mach numbers up to 3 and at altitudes as high as 90,000 ft. Between these and the orbital operating conditions of spacecraft is a huge range of speed and altitude over which airbreathing propulsion should be much more efficient than rocket propulsion. Fundamentally, this is because the airbreathing system draws its oxygen from the atmosphere, while the rocket must carry its oxidizer along. Some idea of the potential of high-speed airbreathing propulsion can be gained from figure 10.1. Values of specific impulse are plotted for a number of engine types over a wide range of M_0; the parameters of each engine are selected as they would optimize for its usual mission. The values for the turbojet, turbofan, and ramjet were taken from chapter 2, those for the supersonic combustion ramjet were estimated by techniques to be explained and are a bit speculative. But the main point, which is the great advantage of airbreathing over rocket propulsion, is indisputable so far as specific impulse is concerned.

This chapter outlines the technical differences between airbreathing engines operating at speeds above $M_0 = 3$ and those for lower speeds. These differences stem mainly from the high stagnation temperature levels which result at large M_0. $T_{t0} = T_0(1 + \frac{1}{2}(\gamma - 1)M_0^2)$ is also shown in figure 10.1. At Mach numbers of 6 and above, conversion of even a part of the air's kinetic energy

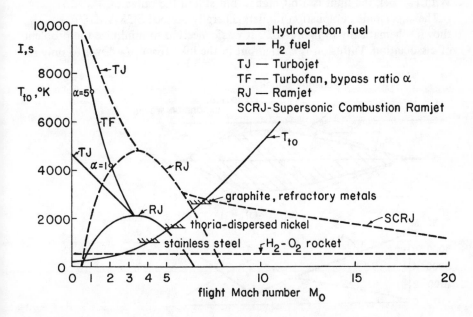

10.1 Specific impulse versus M_0 for various engines, using hydrocarbon and hydrogen fuels.

to thermal energy can raise the temperature to the level where the air dissociates, changing its properties considerably. Even more important, the temperature can be so high that when fuel is added to the air, little temperature rise occurs because what we think of as the normal combustion products, H_2O and CO_2, are strongly dissociated. This occurs above about 2500°K. In addition there are the temperature limitations of structural materials, some of which are indicated in figure 10.1. Above $M_0 = 6$, no material will endure the stagnation conditions, so all structures must be cooled, either radiatively or by the fuel or ablatively.

The factors that limit M_0 are different for the various types of engines. For the turbojet, the limit is set by the turbine temperature limits at about $M_0 = 3$ or 4. With advanced cooling techniques this could be raised, but as shown in figure 10.1 the ramjet is a better choice above $M_0 \approx 3.5$. The H_2-fueled turbojet has a higher I only because the heating value of H_2 per unit mass is some 2.3 times that of jet fuel. Because of the greater cooling capability of the liquid H_2 fuel, the H_2 turbojet could probably be operated to $M_0 = 5$. Conventional ramjets, in which the air is slowed to $M_2 \approx 0.2$ before combustion, as indicated in figure 1.2, operate most efficiently for $2 < M_0 < 6$; above 6 dissociation of the combustion products limits the temperature rise upon which the engine depends for efficient operation. With H_2 fuel, the limit is a bit higher, but still in the range of $M_0 \approx 7$.

The supersonic combustion ramjet, generally dubbed SCRAMJET, is shown schematically in figure 10.2. It was conceived to minimize the problem of dissociation. Diffusion is carried out in the inlet from M_0 down to only

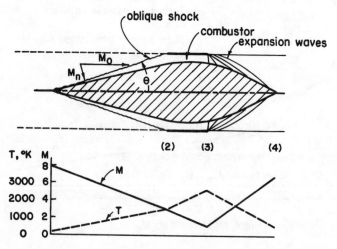

10.2 Schematic of hypersonic SCRAMJET engine (top) showing inlet, combustor and nozzle. Typical Mach number and temperature distributions below.

$M_2 \approx 3$, depending on the value of M_0, so that the air does not reach T_{t0} prior to combustion. The combustion occurs in the supersonic flow, raising T_t and also T, after which the combustion products are expanded through the nozzle. Even though the diffusion is partial, the thermal efficiency can be high because the temperature ratio in the diffuser is still close to 10, giving an ideal Brayton cycle efficiency of 0.90. Combustion in a supersonic flow can be achieved with H_2, though probably not with hydrocarbons, because of the much higher flame speed and wider flammability limits of the H_2.

One of the critical aspects of engines for hypersonic flight is implied by figure 10.2, namely that the engine is mostly inlet and nozzle, and the compression and expansion processes are very critical to the engine's performance. This can be seen more quantitatively by estimating the velocity ratios u_e/u_0 for the engines whose I is given in figure 10.1. We know that

$$I = F/\dot{m}f \approx u_0(u_e/u_0 - 1)/f$$

and that above $M_0 \approx 5$ all engines operate stoichiometrically, so f is a constant that works out to $f = 0.029$ for H_2. Hence

$$u_e/u_0 - 1 = 0.029 Ig/a_0 M_0 \approx 0.00096 I/M_0,$$

which is plotted in figure 10.3; the figure shows that the fractional velocity change across the engine is very small indeed for $M_0 > 6$, so that a small (unexpected) inefficiency in the inlet or nozzle could have large consequences.

These arguments serve to define the critical elements of feasibility for hypersonic airbreathing engines:

1. Cooling by fuel or radiation is required.
2. Inlets and nozzles of high efficiency are essential.
3. Combustion must be carried out at high (supersonic) velocities, instead of at near-stagnation, as in gas turbines.

In the following sections, we will discuss the second and third elements, then close with a few remarks on the first.

10.1 Hypersonic Inlets

The performance of hypersonic inlets is best expressed in terms of a kinetic energy efficiency, which is defined by

$$\eta_d \equiv \frac{\text{available kinetic energy after diffusion}}{\text{available kinetic energy before diffusion}}; \qquad (10.1)$$

it is understood that the flow is expanded to ambient pressure p_0 in both the numerator and denominator. The advantage of η_d is that it is nearly constant over a wide range of M_0.

Now in the notation of figure 10.2 and chapter 2,

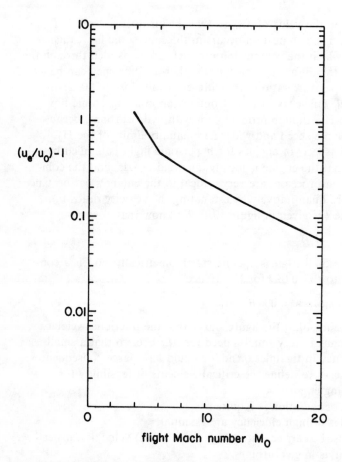

10.3 Fractional velocity change in hypersonic engines.

$$p_{t0} = p_0(1 + \tfrac{1}{2}(\gamma - 1)u_0^2/\gamma RT_0)^{\gamma/(\gamma - 1)}$$

and

$$p_{t2} = p_2(1 + \tfrac{1}{2}(\gamma - 1)u_2^2/\gamma RT_2)^{\gamma/(\gamma - 1)} \equiv p_0(1 + \tfrac{1}{2}(\gamma - 1)\, u_2'^2/\gamma RT_2')^{\gamma/(\gamma - 1)}.$$

The last expression is the definition of the available kinetic energy $u_2'^2/2$. By the definition of η_d, $u_2'^2 = \eta_d u_0^2$, so by conservation of total enthalpy

$$T_2' = T_0 + (u_0^2/2c_p)(1 - \eta_d),$$

and we find that in terms of η_d we can write $\pi_d = p_{t2}/p_{t0}$ as

$$\pi_d = \left[1 + (1 - \eta_d)\left(\frac{\gamma - 1}{2}\right)M_0^2\right]^{-\gamma/(\gamma - 1)}. \tag{10.2}$$

Table 10.1   Values of π_d as a function of M_0.

M_0	1	3	6	10	20
π_d ($\eta_d = 0.97$)	0.97	0.830	0.505	0.193	0.0137
π_d (normal shock)	1.0	0.328	0.0297	3.05×10^{-3}	1.08×10^{-4}
M_n		1.29	1.76	2.48	3.60	7
M_n/M_0		0.59	0.41	0.36	0.35

Given a reasonable η_d (0.97 is a realistic value), we can estimate π_d as a function of M_0 from this relation. Some values are given in table 10.1. Even for such a high efficiency, π_d becomes quite small at large M_0. But the values of π_d for a normal shock at the same M_0 are very much smaller, so the hypersonic inlet must avoid strong shocks in order to achieve $\eta_d = 0.97$. Another way to look at the behavior is to ask what Mach number normal to a single oblique shock is implied by $\eta_d = 0.97$. These values, which are just the normal shock Mach numbers giving $p_{t2}/p_{t0} = \pi_d$, are listed as M_n. Finally, the ratio $M_n/M_0 = \sin \theta$, θ being the wave angle of figure 10.2, is tabulated. The interesting point is that M_n/M_0 is nearly constant above $M_0 = 6$, so an inlet of fixed geometry should be able to operate without much spill with a fixed geometry over this wide range and still yield $\eta_d \approx 0.97$.

10.2   Heat Addition in High-Speed Flow

Next consider the effects of heat addition on the flow in the combustor where $M > 1$. For the sake of simplicity, we will assume that the flow area is constant. Then, in differential form, the equations governing the flow are

conservation of momentum: $(\rho u)(du/dx) = -dp/dx,$ (10.3)

conservation of energy: $(\rho u)/(dh/dx) = u(dp/dx) + dQ/dx,$ (10.4)

conservation of mass flow: $\rho u = \text{const} = \rho_2 u_2,$ (10.5)

where dQ/dx is the rate at which energy is added to the gas by chemical reaction.

To display the fluidynamic effects of heat addition most clearly, assume for the present that $h = c_p T$, c_p (and c_v) constant. Now, from the equation of state $p = \rho R T$,

$$\frac{dp}{dx} = R\left(\rho \frac{dT}{dx} + T \frac{d\rho}{dx} \right) = R\left(\frac{\rho_2 u_2}{u} \frac{dT}{dx} - \frac{\rho_2 u_2 T}{u^2} \frac{du}{dx} \right)$$

where (10.5) has been used twice. Eliminating dp/dx with (10.3), we then have

$$(\rho_2 u_2)u\frac{du}{dx} = -R\left(\rho_2 u_2 \frac{dT}{dx} - \frac{\rho_2 u_2 T}{u}\frac{du}{dx}\right),$$

and from (10.4)

$$\rho_2 u_2 c_p \frac{dT}{dx} = R\left(\rho_2 u_2 \frac{dT}{dx} - \frac{\rho_2 u_2 T}{u}\frac{du}{dx}\right) + \frac{dQ}{dx},$$

which can then be solved for either dT/dx or du/dx. Noting that $M = u/(\gamma RT)^{1/2}$, we find

$$\frac{dT}{dx} = \frac{1 - \gamma M^2}{1 - M^2}\frac{1}{\rho_2 u_2 c_p}\frac{dQ}{dx}, \tag{10.6}$$

$$u\frac{du}{dx} = \frac{(\gamma - 1)M^2}{1 - M^2}\frac{1}{\rho_2 u_2}\frac{dQ}{dx}, \tag{10.7}$$

$$\frac{1}{M}\frac{dM}{dx} = \frac{1 + \gamma M^2}{2(1 - M^2)}\frac{1}{\rho_2 u_2 c_p T}\frac{dQ}{dx}. \tag{10.8}$$

Equation (10.8) shows that heat addition lowers M if $M > 1$ and raises M if $M < 1$, so it always drives M toward 1. Similarly, (10.7) shows that u decreases for $M > 1$ and increases for $M < 1$.

Expressions for the stagnation conditions are more useful than (10.6) and (10.7). From (10.3) and (10.4) directly,

$$\rho_2 u_2 \frac{d}{dx}(c_p T + \tfrac{1}{2}u^2) = \rho_2 u_2 c_p \frac{dT_t}{dx} = \frac{dQ}{dx}. \tag{10.9}$$

The stagnation pressure is $p_t = p(1 + \tfrac{1}{2}(\gamma - 1)M^2)^{\gamma/(\gamma-1)}$, so

$$\frac{1}{p_t}\frac{dp_t}{dx} = \frac{1}{p}\frac{dp}{dx} + \frac{\gamma}{\gamma - 1}\frac{(\gamma - 1)M}{1 + \tfrac{1}{2}(\gamma - 1)M^2}\frac{dM}{dx};$$

using (10.3), (10.7), and (10.8) we find

$$\frac{1}{p_t}\frac{dp_t}{dx} = -\frac{\gamma M^2/2}{\rho_2 u_2 c_p T_t}\frac{dQ}{dx}. \tag{10.10}$$

This equation shows clearly that *heat addition always lowers* p_t; it further shows that the decrease of p_t is much larger for $M > 1$ than for the small values found in gas turbine combustors.

While (10.9) and (10.10) show the effects of heat addition most clearly, it is convenient for purposes of computation to have integrated forms of the equations. Eliminating dQ/dx between (10.8) and (10.7) gives

$$\frac{dp_t}{p_t} = \frac{-\gamma M(1 - M^2)}{(1 + \gamma M^2)(1 + \tfrac{1}{2}(\gamma - 1)M^2)}dM$$

which integrates to

$$\frac{p_{t3}}{p_{t2}} = \left(\frac{1 + \gamma M_2^2}{1 + \gamma M_3^2}\right)\left(\frac{1 + \frac{1}{2}(\gamma - 1)M_3^2}{1 + \frac{1}{2}(\gamma - 1)M_2^2}\right)^{\gamma/(\gamma - 1)}.$$

Similarly,

$$\frac{u_3}{u_2} = \frac{\rho_2}{\rho_3} = \frac{1 + \gamma M_2^2}{1 + \gamma M_3^2}\left(\frac{M_3}{M_2}\right)^2,$$

$$\frac{p_3}{p_2} = \frac{1 + \gamma M_2^2}{1 + \gamma M_3^2},$$

$$\frac{T_2}{T_1} = \left(\frac{1 + \gamma M_2^2}{1 + \gamma M_3^2}\right)^2\left(\frac{M_3}{M_2}\right)^2,$$

$$\frac{T_{t3}}{T_{t2}} = 1 + \frac{Q}{\rho_2 u_2 c_p T_{t2}} = \left(\frac{1 + \gamma M_2^2}{1 + \gamma M_3^2}\right)^2\left(\frac{1 + \frac{1}{2}(\gamma - 1)M_3^2}{1 + \frac{1}{2}(\gamma - 1)M_2^2}\right)\left(\frac{M_3}{M_2}\right)^2. \qquad (10.11)$$

Now, if we fix M_3 at 1, we can give ratios of p_{t3}/p_t^*, T_{t3}/T_t^*, and so forth, as functions of M_2, just as in section 4.1.1. Here we are mainly concerned with the effect of heat addition on p_t, so it is convenient to have p_t/p_t^* and T_t/T_t^* as functions of M. Such are plotted in figure 10.4.

Suppose, for example, that enough fuel is burned to increase T_t by 30 percent, so that from (10.11) we have $Q/\rho_2 u_2 c_p T_{t2} = 0.3$ and $T_{t3}/T_{t2} = 1.3$ in a flow having initial Mach number $M_2 = 3$. From figure 10.4, for $M = 3$ we have $(T_t/T_t^*)_2 = 0.66$ and $(p_t/p_t^*)_2 = 3.4$. Then $(T_t/T_t^*)_3 = 1.3(T_t/T_t^*)_2 = 0.858$, and from the figure $M_3 = 1.7$, $(p_t/p_t^*)_3 = 1.25$, and finally $p_{t3}/p_{t2} = 1.25/3.4 = 0.37$.

10.3   Heat Release Due to Chemical Reactions

Thus far the combustion process has been represented by an external heat source. In fact, no energy is added to the flowing gas mixture during combustion; energy is only converted from chemical to thermal form. The presence of the chemical energy in the unburned mixture can be represented by the addition of a heat of formation $\Delta H_f{}^\circ$ to the enthalpy of each chemical substance other than the elements in their standard forms. The heat of formation is defined as the heat that *must be added* when the elements in their standard states are reacted at standard temperature and pressure (298.16°K and 1 atm) to form the compound in question.

Schematically, in the steady flow process of figure 10.5, if 1 mole of H_2 and 0.5 mole of O_2, both gases at 1 atm and 298.16°K, could be reacted at 298.16°K and 1 atm to form 1 mole of liquid H_2O, the heat that would have to be added would be -68.3174 kcal/mole. Actually, since only the initial

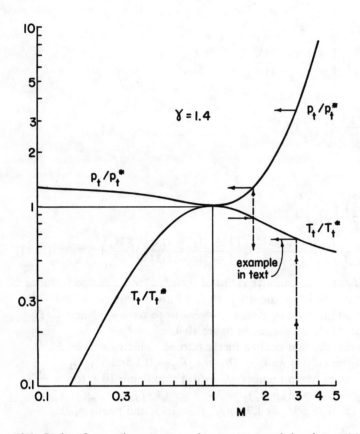

10.4   Ratios of stagnation pressure and temperature to their values at $M = 1$ in flow with heat addition in a constant area duct.

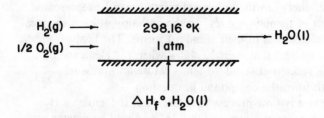

10.5   Schematic of the steady flow process by which the standard heat of formation $\Delta H_f°$ is determined.

and final states matter, the temperature can rise during the combustion process then be dropped by heat transfer out of the flow channel. In fact, this is the way ΔH_f° is measured. Now, if we define a complete enthalpy, say H, by

$$H_i = \int_{T_r}^{T} c_{pi}\, dT + \Delta H_{fi}^{\circ} \tag{10.12}$$

where the reference temperature is usually $T_r = 298.16\,^\circ\text{K}$, then the analogue of (4.2) is

$$\sum x_i H_i + u^2/2 = \text{const}, \tag{10.13}$$

where x_i is the mass fraction of the species i; that is

$$x_i = \rho_i/\rho, \tag{10.14}$$

where ρ is the total gas density and ρ_i is the density of species i. The statement of conservation of momentum remains

$$\rho u\, du/dx = -dp/dx$$

or in integrated form for constant area flow ($\rho u = \rho_2 u_2 = \text{const}$)

$$\rho_2 u_2(u_3 - u_2) = -(p_3 - p_2). \tag{10.15}$$

The equation of state may be written separately for each species,

$$p_i = \rho_i R_i T \tag{10.16}$$

where $R_i = \mathscr{R}/\mathscr{M}_i$, $\mathscr{M}_i$ being the molecular weight and $\mathscr{R}$ the universal gas constant. The total pressure is simply $p = \sum p_i$, and the x_i are related to the p_i by

$$\frac{p_i}{p} = \frac{\rho_i R_i T}{\sum p_i} = \frac{\rho_i R_i T}{\sum \rho_i R_i T} = \frac{\rho_i R_i}{\rho \sum x_i R_i}.$$

Thus, if $R \equiv \sum x_i R_i$, then

$$p_i/p = x_i R_i/R. \tag{10.17}$$

If the mass fractions of the various species x_i were known, then the equations of conservation of mass, momentum, and energy could be solved just as in section 10.2 with the ΔH_{fi}° terms in (10.12) replacing the Q. The x_i are determined by chemical reactions, which may be controlled by kinetic phenomena, as outlined in section 4.4. But in many circumstances the chemical reactions proceed essentially to a condition of local chemical equilibrium. This will happen if the reaction rates are large compared to the rate of the phenomenon that is changing the gas thermodynamic state, usually a flow process in engines.

When the reactions are near equilibrium, the p_i are governed by the law of mass action, which states that for any set of chemically interacting compounds $A, B, \ldots, L, M, \ldots$ that satisfy the stoichiometric equation

$$aA + bB + cC + \cdots \rightleftarrows lL + mM + nN + \cdots \tag{10.18}$$

there is a function K_p of T *alone* such that

$$K_p(T) = \frac{p_L^l p_M^m p_N^n \cdots}{p_A^a p_B^b p_C^c \cdots}. \tag{10.19}$$

The set of equations (10.19), plus the statements that elements are conserved in the chemical reactions, serve to relate the pressure of any chemical species to those of the elements of which it is composed, as functions of the temperature and gas pressure.

For example, suppose H_2 and O_2 react to form H_2O, OH, H_2, O_2, H, and O according to

$$H_2 + \alpha O_2 \rightarrow \beta H_2O + \gamma OH + \delta H_2 + \varepsilon O_2 + \zeta H + \eta O$$

where α is prescribed and we want to find $\beta, \gamma, \ldots$. Usually, α will be prescribed in terms of an "equivalence ratio" ϕ which is the ratio of the fuel flow to that for stoichiometric combustion. Thus, for H_2 and O_2, $\phi = 2/\alpha$. The stoichiometric reactions by which the K_p are defined are

$$H_2 + \tfrac{1}{2}O_2 \rightarrow H_2O \qquad K_{pH_2O} = p_{H_2O}/p_{H_2}p_{O_2}^{1/2} = \frac{\beta}{\delta\varepsilon^{1/2}}\left(\frac{\sigma}{p}\right)^{1/2},$$

$$H_2 + O_2 \rightarrow 2OH \qquad K_{pOH} = p_{OH}^2/p_{H_2}p_{O_2} = \frac{\gamma^2}{\delta\varepsilon},$$

$$H_2 \rightarrow 2H \qquad K_{pH} = p_H^2/p_{H_2} = \frac{\zeta^2}{\delta}\left(\frac{p}{\sigma}\right),$$

$$O_2 \rightarrow 2O \qquad K_{pO} = p_O^2/p_{O_2} = \frac{\eta^2}{\varepsilon}\left(\frac{p}{\sigma}\right),$$

$$\text{(10.20)}$$

where $\sigma = \beta + \gamma + \delta + \varepsilon + \zeta + \eta$. Conservation of elements requires for

$$\text{H:} \quad 2 = 2\beta + \gamma + 2\delta + \zeta,$$
$$\text{O:} \; 2\alpha = \beta + \gamma + 2\varepsilon + \eta. \tag{10.21}$$

Between (10.20) and (10.21) we have six relations from which to solve for the six unknowns $\beta, \gamma, \delta, \varepsilon, \zeta,$ and η, for given values of T and p.

Because the computation is complex, some suggestions as to methods of approach may be in order. The general problem may be posed as follows:

given $\rho_2 u_2$, p_2, T_2, M_2, and the mixture ratio, say of H_2 to air (or equivalent α), find the state of the gas at station 3, that is, p_{t3}, T_{t3}, M_3, and $(x_i)_3$.

1. Consider first the simpler case where $M_2 \ll 1$; then from (10.15), $p_3 \approx p_2$, and in (10.13) we may neglect $u^2/2$ compared to $\sum x_i H_i$. The solution is obtained as follows:

a. Assume a value of T_3 and compute the x_i from (10.20) and (10.21).

b. Compute $(\sum x_i H_i)_3$ and compare to $(\sum x_i H_i)_2$.

c. Iterate the choice of T_3 until (10.13) is satisfied.

The solution for H_2 air is given in figure 10.6.

2. In the general case, where M_2 is not small, we must incorporate the equivalent of the calculation in section 10.2. This can be done as follows:

a. Suppose first that we can specify p_3 rather than p_2. This will be helpful if T_2 is low enough so the $(x_i)_2$ are known independently of p_2. Then we *assume* T_3 and compute the $(x_i)_3$.

b. From (10.13) find u_3.

c. From $\rho_3 u_3 = \rho_2 u_2$ find ρ_2 and hence p_2.

d. From (10.15) find p_2, compare to the value from step c and iterate the choice of T_3.

3. This calculation tends to be tedious. We can simplify it and obtain an approximate solution by using the results of section 10.2 and the fact that the heat release really only depends on the initial and final values of p_2, T_2 and p_3, T_3, which control the gas composition, not on u. Thus the gas velocity

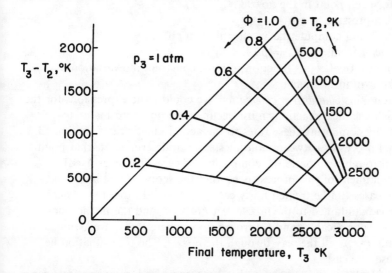

10.6 Temperature rise in combustion of H_2 with air, as a function of final temperature and equivalence ratio, for pressure of 1 atm, and no flow.

is important only in that it lowers T_3 and p_3. Suppose then that we have a solution to problem 1 in the form given in figure 10.6. We can interpret $c_p(T_3 - T_2)$ as an increase in $c_p(T_{t3} - T_{t2}) = Q/\rho_2 u_2$ (see (10.11) because $T_t = T$ for case 1. Then we proceed as follows:

   a. Assume T_3 and read $c_p(T_{t3} - T_{t2})$ from figure 10.6.
   b. Compute $T_{t3}/T_{t2} = 1 + [c_p(T_{t3} - T_{t2})/(c_p T_{t2})]$ and find M_3 from figure 10.4.
   c. Find T_3 from $T_3 = T_{t3}/(1 + \frac{1}{2}(\gamma - 1)M_3^2)$ and compare to the assumed value, iterating choices of T_3 until they are equal.
   d. Obtain p_{t3}/p_{t2} for final M_3 from figure 10.4.

Though approximate because of the assumption of constant values of c_p and γ, this method does include the important phenomena and is much easier than method 2 since the calculations leading to figure 10.6 need be done only once.

10.4 Nozzle and Performance

The flow enters the nozzle in a highly reactive state, and as it expands to lower pressure and temperature, chemical reactions will occur toward completion of combustion, with consequent additional heat release. So long as the expansion is slow enough that near-chemical equilibrium is maintained, the methods given above can be used to compute the flow. We might proceed as follows for each point in the nozzle.

   1. Choose a pressure $p < p_3$.
   2. Assume T and compute the x_i, then u from (10.13).
   3. Compute A from $\rho_3 u_3 A_3 = \rho u A$.
   4. From a plot of A/A_3 as function of p/p_3, construct the variation of p/p_3 for a known nozzle shape.

Having done this *equilibrium calculation*, we could write expressions for the *rates* at which chemical reactions must occur in the nozzle to maintain equilibrium. By comparing these rates to the actual kinetic rates, as limited by collisional processes between molecules, we can determine at what point in the flow through the nozzle the composition freezes (becomes fixed). A detailed treatment of this process is beyond the scope of this book, but two limiting cases can be treated fairly easily: (1) equilibrium flow, where equilibrium is maintained to the nozzle exit pressure, and the method described above is applicable; and (2) frozen flow, where the x_i are assumed to have the values $(x_i)_3$ all the way through the nozzle. The true situation lies between these two cases.

When the ideal nozzle exit velocity has been computed, viscous effects can be accounted for by using either a nozzle velocity coefficient or a kinetic energy efficiency, as defined for the inlet; the former is more usual. It is defined simply as

$$C_u = \frac{\text{actual exhaust velocity}}{\text{ideal exhaust velocity}}$$

and is in the order of 0.98 for a well-designed nozzle at high Reynolds number.

10.5 Cooling

Above $M_0 \approx 6$, the SCRAMJET structure must be cooled, and only two methods are available—radiation to the environment and heat absorption by the fuel as it is consumed. Some very rough order-of-magnitude calculations will suggest the feasibility. First note that if $\dot{Q}$ is the total heat input to the vehicle, $\dot{Q}_r$ is the radiated power, and $\dot{Q}_f$ is the power absorbed by the fuel in being vaporized and heated to the vehicle's surface temperature T_s, then we require $\dot{Q} = \dot{Q}_r + \dot{Q}_f$. Beginning with $\dot{Q}$, we note that if $\dot{q}$ is the heat flux per unit area, then $\dot{q} = St(\rho u c_p T_t)_s \approx (\rho u c_p T_t)_s (C_f/2)$, where C_f is the friction coefficient and the subscript s means evaluated just outside the surface of the vehicle. If the drag $D = \frac{1}{2}\rho_0 u_0^2 C_f S$, where S is the surface area of the vehicle, then

$$\dot{Q} \approx \frac{(\rho u c_p T_t)_s}{\rho_0 u_0^2} D.$$

If the body is black for radiation, $\dot{Q}_r \approx \sigma T_s^4 S$. Let the heat capacity of the fuel, from its liquid state to the temperature T_s be $H(T_s)$; then since the weight flow rate of fuel is F/I, where F is thrust, $\dot{Q}_f = (F/I)H(T_s)$, and we have after dividing by Du_0,

$$\frac{(\rho u c_p T_t)_s}{\rho_0 u_0^3} = \frac{2\sigma T_s^4}{\rho_0 u_0^3 C_f} + \frac{F}{D}\frac{H(T_s)}{u_0 I}. \tag{10.22}$$

To estimate the first quantity, note that from figure 10.2 most of the vehicle's skin is subjected to the conditions after inlet diffusion, so we can estimate the quantities $(\ )_s$ as follows:

$$\frac{p_s}{p_0} \approx \frac{p_{t0}\pi_d}{p_0(1 + \frac{1}{2}(\gamma - 1)M_2^2)^{\gamma/(\gamma - 1)}},$$

$$\frac{T_s}{T_0} \approx \frac{T_{t0}}{T_0(1 + \frac{1}{2}(\gamma - 1)M_2^2)},$$

$$\frac{u_s}{u_0} = \frac{M_2}{M_0}\left(\frac{T_2}{T_0}\right)^{1/2},$$

$$\frac{c_p T_t}{u_0^2} \approx \frac{c_p T_{t0}}{u_0^2} = \frac{1 + \frac{1}{2}(\gamma - 1)M_0^2}{(\gamma - 1)M_0^2}.$$

Collecting these gives

$$\frac{(\rho u c_p T_t)_s}{\rho_0 u_0^3} \approx \pi_d \frac{M_2}{M_0}\left(\frac{1 + \frac{1}{2}(\gamma - 1)M_0^2}{(\gamma - 1)M_0^2}\right)\left[\frac{1 + \frac{1}{2}(\gamma - 1)M_0^2}{1 + \frac{1}{2}(\gamma - 1)M_2^2}\right]^{1/2(\gamma - 1)},$$

and for $M_2 \gg 1$, $M_0 \gg 1$, this is simply

$$\frac{(\rho u c_p T_t)_s}{\rho_0 u_0^3} \approx \frac{1}{2}\pi_d\left(\frac{M_0}{M_2}\right)^{(2-\gamma)/(\gamma-1)} \tag{10.23}$$

The radiative cooling term is

$$\frac{2\sigma T_s^4}{\rho_0 u_0^3 C_f} = \frac{2\sigma T_s^4}{\gamma p_0 a_0 M_0^3 C_f} \approx 1.4 \times 10^{-15}\frac{T_s^4}{(p_0/p_r)M_0^3 C_f}$$

in the stratosphere; T_s is in $°K$ and p_0/p_r is the ratio of local pressure to standard atmospheric pressure. Finally, for liquid H_2 fuel $H(1000°K) \approx 1.9 \times 10^4$ kJoules/kg, so

$$\frac{F}{D}\frac{H(1000°K)}{u_0 I} \approx \frac{F}{D}\frac{6400}{M_0 I},$$

and the heat balance (10.22) is approximately

$$\frac{1}{2}\pi_d\left(\frac{M_0}{M_2}\right)^{(2-\gamma)/(\gamma-1)} \approx 1.4 \times 10^{-15}\frac{T_s^4}{(p_0/p_r)M_0^3 C_f} + \frac{F}{D}\frac{6400}{M_0 I}. \tag{10.24}$$

As an example, take $M_0 = 6$, $M_2 = 3$, $C_f = 0.002$, $T_s = 1000°K$, $\pi_d = 0.5$, $I = 3000$ s; then

$$0.7 \approx \frac{3.2 \times 10^{-3}}{(p_0/p_r)} + 0.36\frac{F}{D}; \quad M_0 = 6.$$

The following *rough* conclusions can be drawn:

1. If the vehicle is to maintain straight and level flight, so $F/D = 1$, some radiative cooling is probably necessary, such that $3.2 \times 10^{-3}/(p_0/p_r) \approx 0.34$ or $p_0/p_r \approx 10^{-2}$, and this implies an altitude of about 100,000 ft or 30 km, which is not unreasonable.

2. If the vehicle is rapidly accelerating, so $F/D > 3$ as for an airbreathing orbital launch booster, then fuel cooling can be sufficient to absorb the entire heat load.

References

10.1 W. H. Avery and G. L. Dugger, "Hypersonic Airbreathing Propulsion," *Astronautics and Space Engineering*, June 1964, pp. 42–47.

10.2 D. P. Hearth and A. E. Preyss, "Hypersonic Technology: Approach to an Expanded Program," *Astronautics & Aeronautics*, Dec. 1976, pp. 20–37.

Problems

10.1 In computing the change in Mach number in an afterburner in section 4.4.4 it was assumed that there was no change in stagnation pressure. Using the method of this chapter, evaluate this assumption and correct the values of M_6 on figure 4.35.

10.2 Generalize the differential argument of section 10.2 to include the possibility of flow area variation, by putting $\rho u A = $ const in place of (10.5). Does this change (10.9)? How is the "thermal choking" condition modified?

10.3 Using the approximate method of section 10.3 to analyze the combustor performance, find the specific impulse of a SCRAMJET operating at $M_0 = 8$ with stoichiometric combustion beginning at $M_2 = 3$. Take $\gamma \approx 1.2$ for the nozzle flow. Compare to figure 10.1.

10.4 A turborocket engine that has been proposed uses hydrazine N_2H_4 as a fuel. The N_2H_4 is chemically unstable, having a heat of formation $\Delta H_f^0 = 12.05$ kcal/mole. It is decomposed in a catalyst bed to produce a hot N_2, H_2 mixture, which is expanded through a turbine. The turbine drives a compressor; its discharge mixes with the turbine exhaust and burns before exiting through the nozzle. Develop expressions for the thrust per unit of airflow and specific impulse of this engine. Plot its performance as a function of M_0, choosing the altitude so that the compressor inlet pressure is constant at 1 atm.

10.5 Following the approximate argument of section 10.5, plot as a function of M_0 the altitude at which a SCRAMJET vehicle must cruise for radiative cooling plus fuel cooling to be sufficient to maintain it in thermal equilibrium. Also, for this altitude schedule plot the diffuser exit pressure p_2.

11
Propulsion Systems Analysis

The problem of propulsion systems analysis is to determine the best propulsion system for some application. Because it generally takes longer to develop the engine than the aircraft, the needs for future aircraft systems must be anticipated by the engine manufacturer. Thus an essential part of engine design and development is the anticipation of future aircraft developments. and the "preliminary design" of engines. On the basis of these preliminary designs, components must be designed and developed long before an aircraft system is committed to design and development.

In the preliminary design phase the criteria for optimization will usually be stated in general terms; for example it may be desired to determine the "best" engine for a next generation commercial transport to serve medium range routes. In this case, best would mean the engine that would lead to the most profitable aircraft while meeting environmental constraints. But the best engine design depends on the aircraft design, so the engine manufacturer must include in his engine studies all those aircraft possibilities most likely to be optimum. Similarly, for a high-performance military interceptor, best might be defined as minimum takeoff gross weight to carry a given payload to some speed and altitude in a given time. Here the optimum altitude versus M_0 trajectory will depend on the engine and vice versa, so an integrated treatment of the engine and airframe is essential.

The first step in attacking such a preliminary design problem is to identify the *figure of merit* by which the system is to be judged. Next, one defines a set of *model propulsion systems* and *model vehicles* whose characteristics can be determined in terms of sets of engine parameters and airframe parameters. A flight plan or mission is chosen, which may also involve parameters to be determined. The problem is then to determine the optimum set of these many parameters.

For these purposes the important performance variables of the engine are thrust/engine weight, thrust/frontal area or thrust/mass flow, and specific impulse. The parameters that might be varied are π_c, θ_t, α, M_T, and perhaps others. The airframe variables might be the lift/drag ratio L/D and the ratio of structural to gross weights W_s/W_g; both depend on many design parameters such as wing loading, aspect ratio, and maximum M_0.

Most aircraft missions can be constructed of the following elements: takeoff, climb and acceleration, cruise, maximum speed, maneuver, loiter, and land. The more critical of these will be discussed briefly.

11.1 Takeoff

During the takeoff roll, assume that $F = F(0)$, that $D = \frac{1}{2}\gamma p_0 M_0^2 A_W C_{D0}$ where A_W is the wing area on which the drag coefficient at zero lift C_{D0} is based. Also assume a rolling friction coefficient C_f. Then, if we neglect the change in mass of the aircraft during takeoff,

$$a_0 m(0) \frac{dM_0}{dt} = F(0) - m(0)gC_f - \tfrac{1}{2}\gamma p_0 A_W C_{D0} M_0^2.$$

Writing this as $dM_0/dt = a - bM_0^2$ where $a = F(0)/a_0 m(0) - gC_f/a_0$, $b = \gamma p_0 A_f C_{D0}/2a_0 m(0)$ and integrating gives

$$M_0 = (a/b)^{1/2} \tanh (ab)^{1/2} t. \tag{11.1}$$

Now, if C_{LT} is the lift coefficient at takeoff, the Mach number at takeoff M_{0T} will be determined by

$$\tfrac{1}{2}\gamma p_0 M_{0T}^2 A_W C_{LT} = m(0)g$$

where A_W is the wing area; inserting this value of M_0 in (11.1) gives the time required for the takeoff roll. More important usually is the length, say X_T, of the takeoff roll, which is

$$X_T = a_0 \int_0^{t_T} M_0 \, dt = a_0 \int_0^{t_T} (a/b)^{1/2} \tanh (ab)^{1/2} t \ dt$$

or

$$X_T = \frac{a_0}{b} \ln [\cosh (ab)^{1/2} t_T] = \frac{-a_0}{2b} \ln (1 - M_{0T}^2 b/a). \tag{11.2}$$

The term

$$\frac{M_{0T}^2 b}{a} = \frac{\gamma p_0 A_W C_{D0} M_{0T}^2}{F(0) - gC_f m(0)}$$

is the ratio of aerodynamic drag at the end of takeoff roll to the net accelerating force at the beginning. There are of course requirements for takeoff with one engine failure, minimum climb rates, and so on which cannot be covered here.

11.2 Climb and Acceleration

For a given airframe and engine, an infinite number of paths in altitude-velocity coordinates can be followed to any desired final altitude and speed. But if a criterion is specified, such as minimum time or minimum fuel usage, then there is an optimum path. It is most easily deduced using the calculus of variations, and the *total energy* formulation of the aircraft dynamic problem (references 11.1 and 11.2).

The total energy E is defined as the sum of the potential and kinetic energies of the aircraft

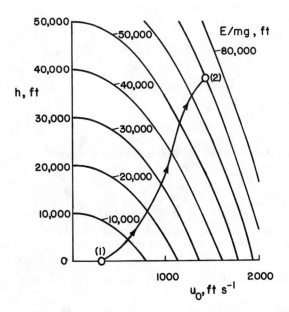

11.1   Total energy E/mg as a function of altitude and flight velocity with possible initial and final points of climb path.

$$E = m(gh + u_0^2/2),\tag{11.3}$$

Its variation with time is due to the net work done on the aircraft by the thrust minus drag, so

$$\frac{d}{dt}\left(\frac{E}{m}\right) = \frac{F - D}{m}\,u_0.\tag{11.4}$$

Contours of constant E/m appear as inverted parabolas on h, u_0 coordinates, as shown in figure 11.1, and the climb and acceleration problem is then to determine the best path for going from a low initial E/m, say at point 1, to a higher value at point 2, according to some criterion.

11.2.1   Minimum Time to Climb

Suppose, for example, that minimum time is desired. From (11.4) the time to climb t_c is

$$t_c = \int_{(E/m)_1}^{(E/m)_2} \frac{m}{(F - D)u_0}\,d\left(\frac{E}{m}\right),\tag{11.5}$$

and the problem is to determine the path on h, u_0 (or h, M_0) coordinates that minimizes the integral. We require knowledge of the integrand as a function of (h, M_0). Since minimum t_c is desired, F should have its maximum value at

any given (h, M_0) (the engine should operate at full throttle), and we can
consider $F = F(h, M_0)$. But D depends on h, M_0, m, and the lift or normal
acceleration, so in complete generality the problem is quite complex. To
illustrate the approach, we may assume the lift equals the weight and estimate
the thrust of an afterburning turbojet from figure 2.5, assuming operation at
constant $W_2 \sqrt{\theta_2}/\delta_2$. The drag coefficient is approximately

$$C_D = C_{D0} + C_L^2/\pi A\!R$$

where C_{D0} is the drag at zero lift, C_L is the lift coefficient, and $A\!R$ is the aspect
ratio of the wing. A typical variation of C_{D0} with M_0 is given in figure 11.2.
If we then choose A_2/A_W (engine size) and $A\!R$, we can plot curves of $(F - D)$
$u_0/m(0)$ as shown in figure 11.3. The vehicle mass m depends on the entire
flight history, but again for simplicity it has been assumed that $m \approx m(0)$,
the initial mass. With these assumptions and approximations,

$$\frac{m(0)}{(F - D)u_0} = f(h, E/m) = g(u_0, E/m);$$

that is, it is a function of h and E/m, and a *different* function of u_0 and E/m.
Let $\eta(E/m)$ be any function of E/m such that $\eta(E/m)_1 = \eta(E/m)_2 = 0$, and
let ε be an arbitrary small constant. Then put

$$t_c(\varepsilon) = \int_{(E/m)_1}^{(E/m)_2} f(H + \varepsilon\eta, E/m)\, d(E/m)$$

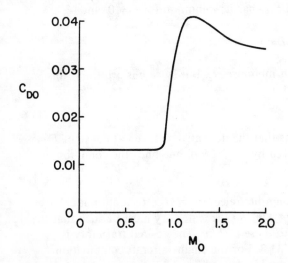

11.2   Typical drag at zero lift of interceptor type aircraft as function of flight Mach
number.

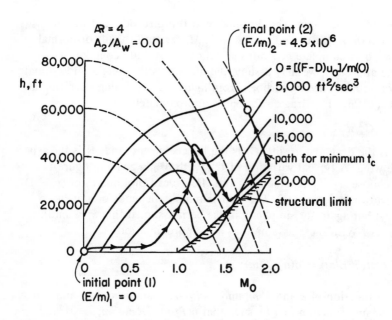

11.3 Acceleration margin as function of altitude and Mach number for interceptor aircraft, showing path for minimum climb time.

where $H = H(E/m)$ is the altitude. Now if H is the desired optimizing function, which minimizes t_c, then $t_c(\varepsilon)$ has its minimum for $\varepsilon = 0$, and

$$\frac{dt_c}{d\varepsilon} = 0 = \int_{(E/m)_1}^{(E/m)_2} \left(\frac{\partial f}{\partial h}\right)_{E/m} \eta \, d(E/m)$$

is the condition on $h(E/m)$ that minimizes t_c. But since η is arbitrary, this condition can only be satisfied if

$$(\partial f/\partial h)_{E/m} = 0 \tag{11.6}$$

where the subscript emphasizes that the derivative is taken at constant E/m. Similarly, if $f(h, E/m)$ is replaced by $g(u_0, E/m)$, one finds the condition

$$(\partial g/\partial u_0)_{E/m} = 0. \tag{11.7}$$

These conditions state that along the trajectory $h(M_0)$, f should not change with h along curves of constant E/m, so the trajectory must be such that the curves of f (or g) are tangent to those of E/m. The desired trajectory is therefore as sketched in figure 11.3. For the engine-aircraft combination depicted, minimum time to climb is achieved by *diving* through the transonic speed range, where the drag coefficient is high, accelerating at low altitude at a speed just below the structural limit of the vehicle (maximum permissible p_{t0}), then "zooming" to the desired final point at constant E/m.

11.2.2   Minimum Fuel to Climb

The fuel consumed in climbing from $(E/m)_1$ to $(E/m)_2$ is

$$m_f = \int_{(E/m)_1}^{(E/m)_2} \frac{dm}{d(E/m)}\, d(E/m) = \int_{(E/m)_1}^{(E/m)_2} \frac{Fm}{I(F - D)u_0}\, d(E/m).$$

Since the minimum fuel consumption probably will not result from full-throttle operation, it is necessary in this case to recognize that the integrand is a function of $\delta = F/F_{max}$ as well as h and u_0. Thus we put

$$\frac{Fm(0)}{I(F - D)u_0} = f(h, E/m, \delta) = g(u_0, E/m, \delta).$$

We must now find the optimum path $H(E/m)$ and the optimum throttle schedule $\mathscr{D}(E/m)$, so we put

$$m_f(\varepsilon_1, \varepsilon_2) = \int_{(E/m)_1}^{(E/m)_2} f(H + \varepsilon_1\eta_1, E/m, \mathscr{D} + \varepsilon_2\eta_2)\, d(E/m)$$

where ε_1 and ε_2 are two arbitrary constants and η_1 and η_2 two arbitrary functions such that $\eta(E/m)_1 = \eta(E/m)_2 = 0$. Since m_f must be a minimum with respect to both ε_1 and ε_2, the conditions for minimum fuel consumption in climb become

$$(\partial f/\partial h)_{E/m} = 0, \quad (\partial f/\partial \delta)_{E/m} = 0;$$

$$(\partial g/\partial u_0)_{E/m} = 0, \quad (\partial g/\partial \delta)_{E/m} = 0. \tag{11.8}$$

The procedure can obviously be extended to allow determination of schedules for any number of variables, but the graphical interpretation becomes a little obscure.

11.3   Cruise

Here the objective may be to attain the maximum range for a given fuel consumption, or it may be to achieve a given range in minimum time. Consider the first case. As before $d(E/m) = ((F - D)/m)u_0\, dt$, but $dm = -(F/gI)\, dt$; so this can be written

$$d(E/m) = -gI\frac{u_0}{m}\, dm - \frac{D}{m}u_0\, dt.$$

The range increment $dR = u_0\, dt$, so that

$$dR = -Iu_0 \frac{gm}{D} \frac{dm}{m} - \frac{m}{D} d(E/m). \tag{11.9}$$

If we neglect the change in E/m, and put $gm = W = L$, this integrates to the classical Breguet range equation:

$$R = Iu_0 \frac{L}{D} \ln \frac{m(0)}{m}; \qquad d(E/m) = 0,$$

which was discussed in section 1.3, but for high-speed flight the change in E/m may be quite important. Furthermore, as the flight speed becomes large, the required lift is reduced by centrifugal force, so that $L = mg - mu_0^2/r$, where r is the radius to the earth's center, and

$$R = -\int_{m_1}^{m_2} Iu_0 \frac{L/D}{1 - (u_0^2/rg)} \frac{dm}{m} - \int_{(E/m)_1}^{(E/m)_2} \frac{L}{gD} \frac{1}{1 - (u_0^2/rg)} d(E/m).$$

In the classical Breguet case, the first integral is maximized by maximizing Iu_0L/D; at hypersonic speeds the optimum u_0 will be larger because the centrifugal lift term increases R. An idea of the magnitude of this effect and of the E/m correction can be had by assuming u_0, L/D, and Iu_0 constant; then

$$R = \frac{Iu_0(L/D)}{1 - (u_0^2/rg)} \ln \frac{m_1}{m_2} - \frac{L/D}{1 - (u_0^2/rg)} \frac{(E/m)_2 - (E/m)_1}{g},$$

where to maintain constant L/D, h must increase from 1 to 2 so that $(p_0)_2/(p_0)_1 = m_2/m_1$. Taking $p_0 \propto e^{-\alpha h}$ gives

$$[(E/m)_2 - (E/m)_1]/g = h_2 - h_1 = \frac{1}{\alpha} \ln (m_1/m_2);$$

finally

$$R = \frac{Iu_0(L/D) \ln (m_1/m_2) - [(L/D) \ln (m_1/m_2)]/\alpha}{1 - (u_0^2/rg)}.$$

From this, if we set m_1/m_2, we can compute R versus u_0, using for example the I dependence of figure 10.1 and estimates for L/D as a function of M_0. Some typical results for $m_1/m_2 = 2$ are as follows:

M_0	$I(S)$	L/D	fuel	R (miles)
0.8	7000	20	JP-4	14,200
2.7	3000	8	JP-4	8,300
12.0	2000	4	H_2	15,300
20.0	1200	3	H_2	16,500

Such results have been interpreted by some to mean that hypersonic transports could be more efficient for long-distance travel than subsonic aircraft.

11.4 Maneuvering

The turning radius of military aircraft in sustained air combat is limited by the thrust because additional drag is encountered due to the lift required for turning. If r is the turning radius, the centrifugal force that must be overcome is mu_0^2/r, so if the turn is in a horizontal plane, the total lift is $L = [(mg)^2 + (mu_0^2/r)^2]^{1/2}$. The ratio of total lift to weight, which is termed the number of g's is then g's $= [1 + (u_0^2/rg)^2]^{1/2}$, or the turning radius is

$$r = (u_0^2/g)/[(g\text{'s})^2 - 1].$$

Some typical values for 2 g's are

M_0	1	2	3	20
r (miles)	3.2	13	29	742

from which it is clear that the pilot of an hypersonic transport will have to plan ahead.

References

11.1 E. S. Rutowski, "Energy Approach to the General Aircraft Performance Problem," *Journal of the Aeronautical Sciences*, Vol. 21, No. 3 (March 1954), pp. 187–195.

11.2 A. E. Bryson, Jr., M. N. Desai, and W. C. Hoffman, "Energy-State Approximation in Performance Optimization of Supersonic Aircraft," *Journal of Aircraft*, Vol. 6, No. 6 (Nov./Dec. 1969), pp. 481–488.

Problems

11.1 Show that if the aerodynamic drag is small compared to the net accelerating force during the takeoff roll, (11.2) reduces to the simple statement that accelerating force times x_T equals vehicle kinetic energy at end of takeoff roll.

11.2 A transport aircraft powered by turbofan engines with $\theta_t = 6$, $\pi_c = 24$ has $L/D = 15$. Estimate the amount of fuel it will use in ratio to its takeoff mass in taking off, climbing to an altitude of 10 km and $M_0 = 0.8$, and cruising on a Breguet path a distance of 4000 km, as a function of bypass ratio α. Use the data of figure 2.6 for the engine performance.

11.3 A fighter aircraft powered by an afterburning turbofan engine with $\alpha = 1$, $\theta_t = 7.5$, $\pi_c = 24$, engages in air combat at about $M_0 = 1$, in which it maneuvers through ten full turns at 4 g's. Estimate the amount of fuel in ratio to the initial total mass consumed during these maneuvers. Assume the aircraft L/D is 10. How far could the aircraft cruise without afterburning at $M_0 = 0.8$ with this same fuel expenditure?

11.4 A transport aircraft is to be powered by either a high-bypass turbofan ($\alpha = 5$) or a turbojet. In either case, the engine is sized by the takeoff requirement that the takeoff roll be 1500 m. Using the simple cycle analysis of chapter 2, estimate the ratios of air mass flow/takeoff mass for the two engine types. Assume $\theta_t = 6$, $\pi_c = 24$ for both. Then estimate the ratio of thrust to maximum available thrust for each engine when cruising at $M_0 = 0.8$, $h = 10$ km. For the takeoff roll, take $C_{D0} = 0.01$, $C_f = 0.02$, $m(0)/A_f \approx 5000$ kg m^{-2}.

11.5 Using the specific impulse values of figure 10.1, compare the ratios of mass in low earth orbit to takeoff mass for a vehicle powered by hydrogen fueled airbreathing engines with that for a two stage H_2–O_2 rocket.

Index

(*Italicized page numbers denote reference to material in figures. Italicized numbers in parentheses denote problem numbers.*)

Accelerating power, 218
Acceleration, *219*
  characteristic time for, 217
  margin for an interceptor, *270*
  stall margin, 219
Acceleration time, 218
  effect of engine size on, 218
Acoustic admittance, 244
Acoustic damping, *243*
Acoustic efficiency, 237
Acoustic equation in duct, 238
Acoustic power, 227
  per unit of jet volume, 232
Activation energy, 99
Actuator discs, 123
Adiabatic film effectiveness, 180, *181*
Adiabatic recovery temperature, 177
Adiabatic wall temperature, 180
Afterburner
  design of, 106–107, 115 (*4.7*)
  fuel systems 219
  length of, 106
  light, 219
  Mach number in, 265 (*10.1*)
  recirculation zone in, 104
  schematic of, *105*
  transients, 219
  of turbojet, *6, 7*
Afterburning low-bypass turbofan engine, *19A*
  specific impulse of, 34, *35*
  thrust of, 34
  thrust per unit of total mass flow, *35*
Afterburning turbojet, *17, 19B*
  compressor pressure ratio of, *28*
  with fixed compressor pressure ratio
    specific impulse of, *31*
    thrust per unit mass flow of, *31*
  specific impulse of, *28*, 29–31
  thrust of, 29–30
  thrust per unit mass flow of, *28*
Aft-fan, *17*
Airbreathing boost to orbit, 274 (*11.5*)
Airbreathing orbital launch booster, 264
Air-cooled turbine, 174, *175*
Aircraft noise, 222
Altitude, effect on afterburner, 115 (*4.9*)
Ambient pressure, effect on thrust, 213

Amplification factor, *199*
Amplitude of vibration, 198
Angular acceleration time, 216
Angular momentum balance, 119
Angular velocity of pressure pattern, 241
Annular burner, 103
Area ratio of nozzle, at separation point, 94
Attenuation
  in duct, 241
  length in duct, 244
  properties of liners, 245
  of sound, 243
Automotive engine, cost limitation of design, 1
Axial flow turbine, 167
Axial velocity, 172
  density ratio, 136
  increments due to blade rows, *127*
  in a turbomachine, 124
Axial vorticity, *142*
Axisymmetric diffuser, 85–86
Axisymmetric mixed-compression inlet, *86*
Axisymmetric throughflow, 123

Base drag, 16, *16*
Bearing load, 201
Bearing stiffness, 201
Bending loads on blades, 193–195
Bending modes of compressor blade, *202*
Bending moment in blade, 193–194
Bending stresses in blade, *193*
Bessel function, 238, *239*
Blade
  cooling, 166
  density, effect on flutter, 206
  flutter, 203–206
  loss, force due to, 207 (*7.4*)
  number, effect on noise, 222
  passing frequencies, 241
  passing tone, 242
  suction and pressure surfaces of, *132*
  tip shock, 242
  vibration, 202–203
  wakes, noise due to, 243
Blades
  axial compressor rotor, 120, *121*
  turbine rotor, 167

Blading
  impulse, *173*, 174
  reaction, *173*, 174
  for subsonic compressor, 129–136
  for supersonic relative velocity, *137*
  transonic, 138
Bleed valve, 213
Blow-in doors, 79, *79*, 97, *97*
Bottoming cycle, 61
Boundary condition for sound in duct, 238
Boundary layer, 73–76
  control, 86
  control in inlet, *86*
  displacement thickness of, 73–75
  effect of pressure on thickness gradient, 75
  on hub and casing, 142
  laminar, 76
  separation, in inlet, 79
  separation point in, *76*
  shock separation of, 76
  suction in inlet, 115 (*4.5*)
  transition, *178*
  turbulent, 76
Brayton cycle
  for ramjet, 5, *5*
  regenerative, *10*
  for turbojet, 6
Breguet range equation, 272
Burner
  combustion inefficiency of, 51
  double-annular type, 111
  efficiency, 103–104
    definition of, 51
  pressure loss in, 51, 103
  of ramjet engine, *4*
  residence time in, 104
  temperature profile in, 103
Burst speed of disc, 191, 206 (*7.1*)
Buzz saw noise, 224, 242, *242*, 243
Bypass doors, *91*
Bypass ratio, 224
  definition of, for turbofan, 31
  effect on thrust and jet noise, *236*, 237
  optimum, 33
  relation to thermal and propulsive
    efficiencies, 20 (*1.6*)

Camber line, 129
Campbell diagram, 202, *203*
Capture streamtube diameter, 80
Capture-to-throat area ratio, *82*

Carbon monoxide as a pollutant, 107–110,
  108
Carnot cycle, 20 (*1.2*)
Carter's rule, 133, *134*
Cascade, 117, *118*, 129
  data, 173
  loss data, *173*
  notation, *130*
  rectilinear, 123
Centrifugal compressor, *156*, 157
  diffuser for, *162*
  efficiency of, 158, 163
  impeller, *159*
  performance map, *162*
  pressure ratio of, *159*, 160
Centrifugal force field, 157
Centrifugal stage, 158
  application as high-pressure compressor,
    158
Centrifugal stress in rotor, 189–190
Certification for noise, 249
Channel flow, 66–68
  definition of, 66
  isentropic, *68*
Chemical reactions, 257–262
Chemical reaction time, 106, *106*
Choked inlet, 224
Choked nozzle, 209–210
Choked turbine characteristic, 185–186
Chord of blade, 121
Climb and acceleration, 268
Combination tone noise, 242
Combined cycle, efficiency of, 63
Combined gas turbine–steam cycle with
  losses, 61–64
Combustion, 100
  efficiency, 110
  incomplete, 107
  in supersonic flow, 253
Combustor, 99
  effiiciency, 111
  operation at idle, 115 (*4.8*)
  of ramjet engine, 4
  turbulent mixing in, 100
Complete enthalpy, defined, 259
Component characteristic, 208
Component matching, 208
Compressible flow in turbine nozzle, 170
Compression
  external, 83–85
  mixed, 85

Compressor, 116–165
axial flow, 116
bleed, 46 (2.2)
characteristic, hysteresis in, 155
design and development, 116
discharge pressure, 213
high pressure ratio, 153
minimum weight, 147, 147
multistage, 151–154, 152, 153
noise blocked by sonic inlet, 223
off-design operation of, 151, 152
performance map, 153
pressure ratio of turbojet
  that maximizes I, 27–28
  optimum, 28
radial flow, 116
spools, 152
stage, design of high-performance transonic,
  164 (5.3)
stall due to acceleration, 219
transonic, 136
-turbine matching, 220 (8.1)
of turbojet, 6, 6
weight, 165 (5.5)
Cone, supersonic flow over, 70
Conical flow fields, 85
Conical shaft, 201
Conservation
of elements in reacting gases, 260
of energy, 259
of momentum, 259
of momentum in direction of blade motion,
  119
of total fluid energy, 119
Continuity equation, 66
Contraction, internal, 85
Control
digital electronic, 215
hydromechanical, 215
logic, 215
surfaces in compressor, 119
system, 213
variables, 213
volume, 14, 15
Convective heat transfer, 12
Convergent-divergent nozzle, 94
Convergent nozzle, 53
  choked, 55
Cooled airfoil, 176
Cooling
of aircraft structure, 47 (2.4)

aspiration, 176
convective, 176
impingement, 176–177
internal, 176–179
internal, of a turbine blade, 178
mass flow, 179
precision required of design, 197
radiative, 263–264
of SCRAMJET, 263–264
Cooling air
effect on drag, 183
pressure loss in, 183
Corrected parameters, 149–151, 164 (5.4)
Corrected speed, 185, 211
definition of, 150
Corrected weight flow, 185, 209, 211
definition of, 150
Cowl drag, 85
Critical speed, 197–200
operation above, 201
operation below, 201
Cruise, 271–272
Cutoff, 240
for fan, 250 (9.6)
tangential Mach number, 240
for wavy wall, 250 (9.1)
Cycle analysis
assumptions in, 21
quantitative, 48–65

Damping
of bearing, 202
of rotor, 201
Decibel, 227–228
equivalent of one atmosphere, 228
Deflection of shaft, 201
Design
of drum rotor, 207 (7.2)
of nozzle vane cooling, 188 (6.5)
of turbine, 187 (6.2)
Deviation angle, 130
D factor correlation, 173
Diffuser, 49, 49, 78–93, 161–163
internal compression type, 114 (4.2)
oblique-shock type, 83–84
pipe, 161
pressure recovery, 49, 49
  definition of, 50
of ramjet engine, 4
starting of, 83
subcritical operation of, 88

Diffuser (continued)
  subsonic, 78–80
  supercritical operation of, 88
  supersonic
    internal compression type, 80–83, *81*
    starting problem of, 81
    throat area of, 82
  vaneless, 161
  variable-geometry type, 91–92
Diffusion, 157–158
  supersonic
    external, 80, 83–84
    internal, 80–83
Diffusion factor, 143–145, *145*, *146*, 173, *173*
  definition of, 131
  for inducer, 160, *161*
  variation with radius, *145*
Diffusion flame, 102
Dipole, 228, *229*
  directional characteristic of, 230
  power of, 228, 230
"Direct" problem, 123
Disc, 189–190
  design, *192*
  rim of, 190
  shape of, 191
  thickness of, 190
  of uniform tensile stress, 190
  vibration, 202–203
Dissociation of combustion products, 23, 252
Distortion
  angle of attack and circumferential, 211
  patterns, *212*
  radial and circumferential, 211
Drag, 3
  coefficient, 266
  definition of, 14–17, *15*
  in supersonic flow, 16, *16*
  at zero lift, *269*
Droplet vaporization, 102
Drum rotor, *176*, 207 (*7.2*)
Duct
  -burning turbofan, 220 (*8.6*)
  linings, 243–245
  modes, 237–241
  propagation, *238*
Duration correction, 248
Dynamics of two-shaft turbofan, 220 (*8.5*)

Effective Perceived Noise Level (EPNL), 245, 246–248
  calculation of, 248
Efficiency, 143–149
  attained by a Carnot cycle, 2
  of centrifugal compressor, 160
  of compressor, definition of, 50–51
  of cooled turbine, 50
    definition of, 183
    degradation due to cooling, 184
  overall, definition of, 1
  polytropic, 51, *52*
  propulsive, definition of, 1–3, *3*
  of stage, 140
  for stage with cascade losses, *141*
  thermal, definition of, 1–2
  of turbine, definition of, 50–51
  of uncooled turbine, 50
Elastic modulus, 196
Energy, converted from chemical to thermal form, 257
Energy exchange
  blade to fluid, 11–12
  kinetic to thermal, 11–12
  rotor to fluid, 116–123
  thermal to surface, 11–12
Energy transfer by wave, 227
Engine
  control variable, 210
  dynamics, 215
  -inlet matching, 211
  pressure ratio, 20 (*1.3*), 209
  temperature ratio, 20 (*1.3*), *209*
  time scales in, 215
Entropy
  change in stage related to heat transfer, 183
  disturbance, 226
  increase by transfer of heat to cooling flow, 183
  rise in blading, 138–139
Environmental Protection Agency 1979 proposed emissions limits, 109
Equation of state, 259
Equilibrium constant, 260
Equivalence ratio, 100, *106*, 260
  definition of, 100
Euler equation, *118*, *119*, 119–120, 143, 164 (*5.2*), 167
  for incompressible fluid, 120
External compression variable-geometry inlet, 92, *92*

Fan
  front, *18*
  noise, 223
  pressure ratio of, condition on, 34
  stage, 150–151, *150*
  of a turbofan, 8, *8*
Federal Aviation Rule, Part 36 (FAR-36),
  222, 245
Figure of merit, 266
Film
  coefficient, 180
  cooling, 102, 176, 188 (*6.4*)
    effectiveness of, 180
    effect on boundary layer, 179
    effect on heat flux, *180*
    stress due to, 207 (*7.3*)
  -cooling airflow, effect on kinetic energy
    efficiency of turbine cascade, *185*
  and transpiration cooling, 179
Flameholder, 104, *105*
  dimension, 107
Flame speed, 101, *101*
  laminar, 101, *101*
  turbulent, 101
Flame stabilization, 106
Flow
  equilibrium, 262
  frozen, 262
Fluid density, effect on flutter, 206
Flutter
  bending, 205
  influence of neighboring blades, 204
  instabilities, 203
  stalled, 204, *204*
  unstalled, 204
Forced oscillation, 198
Frequency spectrum of jet noise, 231, 232, *233*
Friction coefficient, 77, 177
Fuel
  energy content of, 4
  minimum to climb, 271
  use by transport aircraft, 273 (*11.2*)
  use in air combat, 273 (*11.3*)
Full stall, 155

Gas generator, *6*, 7, *17*, 208–209, 212
Gas properties, variation of, due to
  temperature changes, 48
Gas turbine combined cycle, 42
  power plant, *10*, 11
  thermal efficiency of, 42, *43*

Gas turbine–steam combined cycle
  efficiency of, *61*, *64*
  specific power of, *61*
Gas turbine topping cycle, 42
Gas turbine with bottoming cycle
  efficiency of, *43*
  specific work of, *43*
Guide vane and stator angles, 213

Heat absorption by fuel, 263
Heat addition
  effect on Mach number, 256
  effect on stagnation pressure, 256, *258*
  with flow area variation, 265 (*10.2*)
  in high-speed flow, 255–257
Heat engines, definition of, 1
Heat exchanger
  effectiveness of, 53
  pressure drop in, 114 (*4.1*)
Heat flux, 180
Heating value of fuel, 23
  definition of, 51
  lower, 52
  upper, 52
Heat of formation, *258*
  definition of, 257
Heat transfer
  to blade, estimated, 178
  on cooled turbine blade, *178*
  to cooling fluid, 178
  gas-solid, 76–78
  rate, 181
  to surface, 176
  to turbine nozzle vane, 188 (*6.3*)
Helicopter, 164 (*5.2*)
  tip jet drive for, 164 (*5.2*)
Helmholtz resonator, 243
High-bypass turbofan engine, *19*
Hot gas ingestion, 219
Hub and casing, cross section of, 164 (*5.3*)
Human response to noise, 245
Hydrazine fuel, 265 (*10.4*)
Hydrogen, 100–101, *101*
Hydrogen-fueled turbojet, 252
Hypersonic airbreathing engine, critical
  elements of, 253
Hypersonic engine, 251–253
  fractional velocity change in, *254*
Hypersonic inlet
  of fixed geometry, 255
  pressure recovery for, 254–255

Hypersonic ramjet, performance of, 262

Ideally expanded nozzle, 55
Ideal nozzle, 53
Ideal ramjet
  specific impulse of, 22
  thrust of, 22
Impeller, *156*, 157
  backward curved, 165 (*5.8*)
Imperfect expansion, 57
  of nozzle flow, 52–53
Impulse stage, *168*
Impulse turbine, 167, 169
Incidence, 130
  minimum loss, 130
Inducer, *156*
Inlet
  off-design behavior of, 86
  of ramjet engine, 4
  start, *91*
  unstart, 114 (*4.4*)
  variable-geometry, 211
Inlet distortion, 211
Inlet guide vanes, 120, *121*, 125, 241
  effect of, 148–149
  effect on pressure ratio and efficiency, *149*
  effect on relative Mach number, 129
Interblade phase angle, 204–205
Internal compression diffuser, off-design
  behavior of, *88, 89*
Internal compression inlet, fixed geometry,
  82
"Inverse" problem, 123
Isentropic diffuser, 84
Isentropic flow, 67
Isentropic spike inlet, *84*
Isothermal effectiveness, 181, *182*

Jet
  instability of, 230
  mixing layer of, *230*
Jet acoustic efficiency, 231
Jet acoustic power, 231, *234*
Jet noise
  directional characteristic of, 231, *232*
  reduction vs thrust, *236*
  suppressors, 222, 234
  of turbofan engine, 222
  of turbojet engine, 250 (*9.4*)
  variation with bypass ratio, 250 (*9.5*)

Kinetic energy efficiency, definition of, 184,
  253

Law of mass action, 260
Leakage
  effect on efficiency, 172
  in turbines, 172
Lip of inlet, 79, *83, 84*
Logarithmic decrement, 205
Loss factor, 133, 144
  and compressor efficiency, 138–140
  correlation for, 133
  definition of, 139
Loss parameter, 144
  correlation with diffusion factor, *134*

Mach number
  in afterburner, *105*
  axial flow, 122, 149
  of compressor or fan, 78
  critical value of, 82
  definition of, 11, 22
  effect on losses, 131, *131*
  flight, 6
  limits, 252
  maximum, *171*
  of piston motion, 240 (*1.4*)
  ratio for shock on wedge, *71*
  relative, 144
    in compressor, 136
    effect on stage efficiency, 140
    at stator entrance, 159
  surface, for flow over cone, *72*
  tangential, 122, 149, 239
    effect on pressure ratio and stage
      efficiency, *148*
  for turbine blades, 166
Maneuvering, 273
Mass flow, 143–147, *171*
  capacity, 158, 160, *161*
  per unit of engine frontal area, 145
  ratio, 180
  vs pressure ratio, 145–147
  vs pressure ratio characteristics for a
    compressor, 153, *153*
Mass fraction, 259
Mechanical damping, 205
Minimum climb time, trajectory for, *270*
Minimum loss incidence, definition of, 131
Mixed-compression inlet, *85*, 114 (*4.3*)
  variable-geometry type, *91*

Model propulsion systems, 266
Model vehicles, 266
Modes of vibration of blades, 202
Moment of inertia of rotor, 216
Momentum equation in compressor, 117
Momentum transport, 76
Monopole, 228, *229*
   power of, 228

Nacelle installation, of turbofan engine, *19*
Nitrogen oxides
   impact on the ozone of the stratosphere,
      111
   as pollutants, 107–113
   production by combustors, *112*
NO
   effect of combustor inlet temperature on,
      111
   mass fraction of, *110*
   production of, 108–113, *108*
Noise, 151
   vs aircraft size, 250 (*9.7*)
   compressor, propagation through inlet, *92*
   Effectiveness Forecast (NEF), 245–247
      for Kennedy Airport, *246*
   fan, 80
   generation, 149
   jet, 230–234
   limits imposed by FAR-36, *248*, 249
   masking of, 248
   measuring stations, 222, *222*
   quantitative indices of, 245
   reduction
      due to bypass ratio, 235
      due to ideal ejector, 235
   restrictions, 222
   due to shock, 233, 234
   sources, 13–14, *223*, 224–226
   of subsonic engines, 223, *223*
   of supersonic engines, 223, *223*
   supersonic jet, 232, *233*
   suppression, 234–237
      due to opening nozzle, 235
   takeoff, 224
      landing and sideline, 222
   from turbomachinery, 237
Noisiness of an individual aircraft, 245
NOYs, 247, *247*
Nozzle
   area, 220 (*8.2*)
   area ratio, 93

base drag of, 96
ejector type, 97–98, *97*, *99*
exhaust, 93–99
fixed, 94–96
flow, 94
losses, 48
matching, 210
optimum, 171
pressure ratio, 93, 212
of ramjet engine, 4, *4*
throat area
   and area ratio, *94*
   to compressor inlet area, 93
thrust of, 95
turbine, 167

Oblique-shock inlet, *83*
Oil damped bearings, 202
Operating limits of engine, 215
Operating line, 209, 210, 218
Operating range for ramjets, 23
Optimization criteria, 266
Optimum climb path, 267
Optimum compressor pressure ratio, with
   afterburning, *28*, 29
Overexpanded nozzle, *95*, 96
Overtemperature, compressor stall due to,
   210

Partial stall, 155
Perceived noisiness, 247, *247*
Performance estimation, 213
Performance maps for fan stages, *150*
Performance of engine, 208
Phase lag in surge, *199*
Plane waves, 227
Plug type nozzle, *96*
Point source of sound, 227
Pollutant emissions, 109
Pollutant formation, 107–113
Potential vortices, 124, 125
Power balance of compressor-turbine, 26,
   54–55
Power lever, 215
Power matching, 216
Prandtl number, 12, 78
Preliminary design, 266
Pressure ratio, 129, 158
   limited by diffusion factor, *146*
Pressure recovery, 80, 83, 88, 90, *90*
   best wedge angle for, 84

Primary combustor, 102–104, *102*
Primary jet of turbofan, 8, *8*
Primary zone residence time, 109
Principal axes, 193
Profile losses in a turbine, 173
Propagating stall, 154–155
Propagation of sound from inlet, 237
Propane, 101, *101*
Propulsion system, 208
Propulsive efficiency, 223
  of turbofan, 32, 45, *45*
  of turbojet, 7
Pumping characteristics, *209*, 210, 212, 220
  (*8.1*)
Pure tone sound, 224, 247

Quadrupole, 228, *229*
  directional radiation characteristics of, 250
  (*9.2*)
  power of, 230

Radial equilibrium, 124, 128
Radial flow turbine, 187 (*6.1*)
Radial motion in compressor, 139
Radial pressure gradient, 124
Ramjet, 4–6
  with losses, 65 (*3.1*)
  thrust of, 46 (*2.1*)
Range of aircraft, 271
  definition of, 3–4
  effect of centrifugal force on, 272
  for high Mach numbers, 272
Reaction
  definition of, 169
  degree of, 167, 171, 188 (*6.6*)
  50%, 169
  radial variation of, 171–172
  rate, 99, 100, 259
  rate of chemical, in nozzle, 262
Recirculation zone, 102
Reduced frequency, 206
Reduced velocity, definition of, 206
Regenerated engine, efficiency of, 10
Regenerated gas turbine
  efficiency of, *61*
  with losses
    effect of turbine inlet temperature on
      efficiency, *60*
    effect of turbine inlet temperature on
      specific power, *60*
  specific power of, *61*

Regeneration, 8–11, *9*, *10*
Regenerative engine
  efficiency of, 40
  tradeoff between power and efficiency of, 41
Regenerative gas turbine, 39–41
  with losses, 57–61
    thermal efficiency of, 59, *60*
    specific fuel consumption of, 40
    thermal efficiency of, 40
Regenerator, *9*, 39
  definition of effectiveness, 53
  effectiveness of, 39, 59
  pressure losses in, 53, 59
Relative Mach number, for excitation of
  propagating mode, 240
Residence time, 106
Resonance, 198–199
Resonant excitation of blade, *203*
Resonant frequency, 244
Resonant speed, 199
Resonators, *243*
Reynolds analogy, 77, *77*, 78
Reynolds number, 12, 76, 150, 177
  effect on deviation angle, *135*
  effect on loss parameter, 134, *135*
  magnitude of, 136
Rocket propulsion compared to airbreathing
  engines, 251
Rotating stall, *154*, 157
Rotational energy of the rotor, 218
Rotor, *156*
Rotor-relative temperature, 188 (*6.6*)
Rotor stagnation temperature, related to
  reaction, 169
Rotor-stator interaction, noise due to, 241
Rotor wakes, 241

SCRAMJET, 252, *252*, 253
  altitude of operation, 265 (*10.5*)
  compression and expansion processes in,
    253
  specific impulse of, 265 (*10.3*)
  thermal efficiency of, 253
Screens, 211
Secondary air, 92
  in nozzle, 97
Separated flow
  in a subsonic channel, *74*
  on suction surface of blade, *74*
Separation, 73–76, 166
Separation point in nozzle, 94

Shaft critical, 201
Shaft deflections, 200, *200*
Shaft engine, 8–11, *9*
Shaft turbine
  specific fuel consumption of, *41*
  specific work of, *41*
Shear stress, 77
Shock
  detached, 70
  in a divergent passage, 87
  normal, 69
  oblique, 70
  stagnation pressure ratio across, 69, *69*
  static pressure ratio across, 69, *69*
  upstream of blades in a compressor, 137
Shock-boundary layer interaction, 95
Shock-boundary layer losses, 173
Shock losses, 120, 122, 144, 148
  in the rotor, 147
  in the stator, 148
  in turbine blades, 174
Shock waves, 68–73
  in supersonic inlet, 49
Shrouded turbine, 172
Shroud or casing, *176*, 203
Sideline noise, 224
"Slip," 160
"Slip" factor, definition of, 160–161
Small disturbances in a fluid, 226
Solidity, 122
Soot formation as a pollutant, 107
Sound, ear's response to, 227
Sound power, 227
  level, 228
Sound pressure level, 228
Sound radiated by airfoil, 250 (*9.3*)
Sound velocity, 226
Spacing, blade, 121
Specific impulse, 20 (*1.2*), 213, 220 (*8.3*)
  definition of, 3–4
  of a ramjet, 23
  of a simple ramjet, *24*
  of a simple turbojet, *28*, *31*
  of a turbofan, 32
  of a turbojet, equations for, 26–27
  for turbojet with losses, 55–56
  for various engine types, 251, *251*
Spill
  subsonic, *89*
  supersonic, *89*
Spillage drag, 16, *16*

Spill of excess mass flow, 82
Spill shock, 87
Stability in flutter, 206
Stage
  definition of, 120
  efficiency
    definition of, 139
    related to loss factors, 140
  number required as function of tangential
    Mach number, 165 (*5.6*)
  pressure ratio, 123, 143–149
  temperature ratio, 120–123
  velocity diagram, *121*
Staging, 128
Stagnation enthalpy, 67
Stagnation pressure, 22
  of compressor of a turbojet, 25
  definition of, 22
  divided by ambient pressure, 22
  loss due to shocks, 137
  ratio, 22
    for a compressor rotor, 163 (*5.1*)
    for a normal shock, 87
Stagnation temperature, 6, 22, 117, 251
  definition of, 22
  divided by ambient temperature, 22
  ratio, 22
    for a compressor rotor, 163 (*5.1*)
Stall, 152, 219
  in compressors, 154–157
  line, 151, 212
  margin, 211, 218–219
  prediction, 211
Stanton number, 77, 176, 177
Static pressure ratio in impeller, 157
Stationary gas turbines, *10*, 11
Stator blades, 120, *121*
  variable-angle type, 165 (*5.7*)
Stiff rotor, 201
Stoichiometric fuel-air ratio, 100
Stoichiometric ramjet
  specific impulse of, 23
  thrust of, 23
Stoichiometric reaction, 260
Stratosphere, 112
Streamtube, 119, *119*
  centrifugal force and radial acceleration of,
    124
Stress, 13, *13*
  in blades, 192
  due to centrifugal forces, 13, *13*

Stress (continued)
  in discs, 151
  in thin rotating ring, 20 (*1.5*)
  in turbomachine rotor, *189*
Strouhal number, 232
Subsonic diffuser
  schematic of, *79*
  suction in, *86*
Subsonic diffusion, *73*
Subsonic jets, 231
Substantial derivative, 225
Sulfur, in aircraft fuels, as a pollutant, 107
Supersonic combustion ramjet, 251
Supersonic diffusers, 80
Supersonic diffusion, 80
Supersonic transport, 224
Surge, 155
  margin, 216
  mechanism, *156*
Sweep, 163
Swirl
  exit, 172
  in exhaust gas, 167
Systems analysis, 266

Takeoff, 266–267
  roll, 267, 273 (*11.1*)
Tangential Mach number, 122, 149, 239
  variation of pressure ratio and efficiency
    with, 147
Temperature ratios, of compressor of
  turbojet, 25
Temperature rise
  in combustion of H_2, 261
  dependence on radius, 125
Tertiary air, 97, *97*
Thermal choking, 104
Thermal coefficient of expansion, 195
Thermal conduction, 177
Thermal conductivity, 188 (*6.3*)
Thermal efficiency of ramjet, 6, 20 (*1.1*)
Thermal efficiency of turbojet, 6
Thermal stress, 177, 195–197
  in blades, *196*
  related to heat flux, 196–197
Thermal transport, 76
Thermodynamics, First Law of, 116
Thickness of blade skin, 197
Three-dimensional flow, 140, 142, *142*
Throat of inlet, 87
Thrust, 212, 220 (*8.3*)

control by opening nozzle, 220 (*8.4*)
  definition of, 2, 14–16, *15*
  equation, 15
  of JT3D, *214*
  lapse, 274 (*11.4*)
  level, 215
  loss due to noise suppression, 235, 237
  of ramjet, 20 (*1.1*)
    as function of Mach number, 24, *25*
  of turbofan, 32
  of turbojet, equations for, 26
Thrust per unit mass flow
  of simple ramjet, *24*
  of simple turbojet, *28*
    with fixed compressor ratio, *31*
Time response of the corrected speed, *217*
Time to climb, 268–270
Tip clearance leakage, 140, 142
Tip clearance losses, 151
Tone correction, 248
  to perceived noise level, 247
Torque, 119
Total energy
  of aircraft, *268*
  definition of, 267
  equation, 117
  formulation of aircraft dynamics, 267
Total pressure, 144
  loss coefficient, definition of, 130
Transition, 177
Turbine, 166–186
  characteristics, 186
  -compressor matching, 166
  cooling, 172
  disc, *176*
  efficiency, 166
  efficiency correlations, 172
  life, effect of temperature on, 197
  nozzle area, 210
  outlet flow area, 167, *168*
  outlet temperature, 215
  performance map, *186*
  shaft, 208
  stages, number of, 166
Turbine inlet temperature, 43–46, 166, 174,
  175
  effect on efficiency
    of gas turbine combined cycle, *46*
    of regenerated gas turbine, *46*
  effect on specific impulse
    for simple turbojet, 44, *44*

Turbine inlet temperature (continued)
  of turbofan engine, *45*
  effect on thrust for simple turbojet, 44, *44*
  limit, 197
  limited by thermal stress, 178
  of turbojet, 7
Turbofan, 7–8, *8*, *18*, 208
  dual nozzle, 65 (*3.2*)
  ductburning in, 47 (*2.8*)
  engine
    effect of bypass ratio on, *33*
    required fan pressure ratio for, *33*
  with equal jet velocities, 32, *33*
  nonafterburning
    specific impulse of, *35*
    thrust per unit of total mass flow of, *35*
  specific impulse of, 45
Turbojet, 6–7, 208
  cycle analysis of, 25–28
  cycle
    effects of compressor efficiency on, 56–57
    effects of pressure losses on, 56–57
    effects of turbine efficiency on, 56–57
  engines, effect of compressor pressure ratio
    on, *30*
  with losses, 54–56
    specific impulse of, *58*
    thrust of, *58*
  thrust of, 47 (*2.5*)
Turbomachinery noise, 222
Turboprop
  distribution of power between jet and
    propeller of, 38–39
  engine
    propulsive work as function of compressor
      pressure ratio, *38*
    specific fuel consumption of, *38*
  fuel consumption of, 37
  propeller of, 36
  propulsive efficiency of, 36
  propulsive power of, 36–37
  specific fuel consumption of, 37
Turborocket engine, 265 (*10.4*)
Turbulence, represented by quadrupole, *229*
Turbulence noise, *225*
Turning radius, 273
  for hypersonic transport, 273
Two-dimensional flow, applicability in
  compressor blading, 136
Two-spool turbojet engine, *18*

Unburned hydrocarbon as a pollutant, 107,
  109
Underexpanded nozzle, *95*
Unstable inlet, 87
Unstarts, 89
Unsteady compression, 117
Unsteady distortion, 212
Unsteady flow noise, *225*

Variable cycle engine, 224
Variable-geometry nozzle, 96–99
Variable stator, 152
Velocity coefficient, 262–263
Velocity diagram, *118*, 120, *121*
Velocity ratio, 253
Velocity triangle, *126*
  for a turbine stage, *168*
Vertical takeoff engine design, 165 (*5.5*)
Vibrating blade, represented by a dipole,
  229
Vibration, 197–200
Vibrational model of turbomachine rotor,
  198
Vibration excitation, 202
Vortex design of compressor, 125
Vorticity disturbance, 226

Wake momentum thickness, 132, *133*
Wake structure due to swirl, 243
Wave angle
  for flow over cone, *72*
  for oblique shock on wedge, *71*
Wave equation, 226–227
Wave number, 239
Wave-off from aircraft carrier, 220 (*8.4*)
Wedge, supersonic flow over, *70*
"Windmilling," 152, *152*
Work coefficient, 143–149, *145*, *146*, *147*
Work done on blade, 205
Work factor limited by diffusion factor, *146*